Superconductivity Revisited

RALPH DOUGHERTY
J. DANIEL KIMEL

CRC Press
Taylor & Francis Group
Boca Raton London New York

CRC Press is an imprint of the
Taylor & Francis Group, an **informa** business

CRC Press
Taylor & Francis Group
6000 Broken Sound Parkway NW, Suite 300
Boca Raton, FL 33487-2742

First issued in paperback 2019

ISBN-13: 978-1-4398-7426-4 (hbk)
ISBN-13: 978-0-367-38061-8 (pbk)

Library of Congress Cataloging-in-Publication Data

Dougherty, Ralph C., 1940-
Superconductivity revisited / Ralph Dougherty, J. Daniel Kimel.
p. cm.
Includes bibliographical references and index.
ISBN 978-1-4398-7426-4 (hardback)
1. Superconductivity. I. Kimel, J. Daniel. II. Title.

QC611.92.D68 2012
537.6'23--dc23   2012035069

Visit the Taylor & Francis Web site at
http://www.taylorandfrancis.com

and the CRC Press Web site at
http://www.crcpress.com

Contents

Preface

In 2002 one of us, Ralph Dougherty, was asked to put some energy into the second semester course in freshman chemistry at Florida State University. The text that had been selected for the course dealt with the subject of superconductivity, though the mechanism of superconductivity was skillfully sidestepped. This book is a broadly edited and expanded version of the material that was shared with the freshman chemistry students on the conduction of electricity without resistive energy loss—superconductivity. Lectures on superconductivity were developed after about two weeks of intense reflection on the subject from the perspective of a physical-organic trained chemist. Shortly after the lectures on superconductivity, we looked at the original literature in physics to evaluate the current status of the field and found that current theory did not involve the concept of reduction of electron scattering associated with nodes in the wave function for the conductor that included the atomic nuclei in the lattice, as was suggested to the students.

When a physical chemist walks into condensed matter physics without any formal introduction, it is a little like exchanging glasses with someone whose prescription is about five diopters different from yours. In the strain to recognize anything, it is sometimes possible to see new things. Not all new things are useful, and it takes some filtering. We have benefited from the fact that cycling, both on and off road, is one of the main sources of exercise for the two authors.

When the two of us started on this book in 2010, we agreed to have a serious look at areas that might be susceptible to critical reevaluation in the foundations for superconductivity. The physics foundation for superconductivity is condensed matter physics. The first area that came up was the concept of *magnetoresistance*.[*] Anisotropic magnetoresistance, which involves conduction in ferromagnets, was discovered in 1856 by William Thompson, later known as Lord Kelvin. *Magnetoresistance*, as it generally appears in condensed matter physics, was first observed by Edwin Hall in his experiments on the Hall effect in 1879.

Hall resistance, ρ_{xy}, in the Hall effect, is an expression of magnetoresistance; that is, resistance to the movement of electrons caused by an external magnetic field. The concept of electron "effective mass" in electron cyclotron resonance also appears to be a result of magnetoresistance. Some details related to the relationship between electron effective mass and magnetoresistance can be seen in Chapter 10.

We are able to show that positive Hall coefficients can arise as a result of angular momentum states of low momentum electrons at the ground state Landau level. The negative z-component of angular momentum electrons that populate this state nearly exclusively appear as the source of the electron negative current on the y axis in the Hall experiment that results in a positive Hall coefficient.

---

[*] If you are a novice, technical terms like *magnetoresistance* or *Hall resistance* can be found in the glossary and the index. At this point you might just read these terms as words. You will understand their meanings as the book develops.

The Wiedemann–Franz law is the empirical relationship between thermal and electrical conductivity in metals. Sommerfeld placed this law on a quantum mechanical, electronic basis in 1927 [1]. The Sommerfeld relationship also strongly suggests that the quantum mechanical, electron scattering process in metals is an exclusively electronic process (see the method of partial waves [2]). Partial wave scattering as the source of electronic resistivity in metals is one of the central themes of this book. When conditions of temperature and pressure are such that partial wave scattering of conducting electrons vanishes, the system undergoes a phase transition and becomes a superconductor, if it is physically large enough.* The energy gap in a superconductor is the gap between the energy of the Fermi level for the superconductor and the Fermi level for the normal conductor state of the same material at the same temperature.

A second major theme of this book is the importance of electron magnetic coupling in conductivity. In metals, when electrons become spin paired in a single conduction band wave function, their mobility universally decreases. They are on the path to being insulators. This can be seen in the near-Mott transition between tetragonal, distorted face-centered cubic tin metal and diamond lattice tin, the semiconductor. Superconductors do not escape the effective reduction of inter-electron repulsion that is associated with magnetic coupling in spin pairing. This effective reduction in inter-electron repulsion is responsible for the existence of the electron pair bonds, which are the foundation of modern chemistry and materials science. The bonding-ness of spin-paired electrons will always decrease mobility of the electrons in a neutral lattice of metal atoms.

Magnetic coupling of unpaired electrons at the valence level in metals with unpaired inner shell electrons is exceptionally important to the strength of metallic bonds. It appears that metallic, single electron bonds are the strongest chemical bonds as judged by the one bar vaporization temperatures of refractory elements.† The two top elemental superconductors, niobium, Nb, and technetium, Tc, are seventh and fifth, respectively, in the list of refractory elements (Table 7.2). In electron pair bonding, the paired electrons' repulsion is effectively diminished by the magnetic coupling enthalpy associated with spin pairing. In metallic bonds, exothermic magnetic coupling between the bonding electron and nonbonding electrons on the bonded atoms plays a similar role in effectively increasing the coulombic binding energy. Because of the increased coulombic attraction when magnetic couplings also operate, the strongest chemical bonds are seen with the metals that have the maximum potential for electron–electron magnetic coupling using partially filled d subshells.

A third theme of this book is magnetic susceptibility. At the superconductivity phase transition, the volume magnetic susceptibility of the metal phase discontinuously changes from the order of $\pm10^{-5}$ for a normal metal to -1 for a superconductor. Diamagnetic metals‡ such as lead, and paramagnetic metals† such as niobium, both form superconductors. In superconductivity the phase transition that changes

---

* Superconductors must be three-dimensional and large enough to have a non-superconducting layer, λ_L. Superconductors support a volume magnetic susceptibility of -1 for the body of the phase.
† See Chapter 7, Table 7.2.
‡ Diamagnetic materials have susceptibility $0 > \chi \gg -1$. Paramagnetic materials have susceptibility $0 < \chi \ll 1$.

the magnetic susceptibility occurs at a critical point. The magnetic susceptibility changes from paramagnetic or diamagnetic, $|\chi| \approx 10^{-5}$, to pure diamagnetic, $\chi = -1$. The fundamental cause of this change in magnetic susceptibility is the quantum controlled end of dissipative electron scattering in the conduction band. This single event prepares both phases for the phase transition. In the normal metal, loss of dissipative electron scattering appears to collapse the definition of the Condon magnetic domains in the metal (see Chapter 9). On the superconductor side of the transition, the end of dissipative electron scattering defines the superconductor body as a single magnetic domain. If an external magnetic field approaches or is present in the newly formed superconductor it will generate an opposing diamagnetic current in the superconductor—Faraday's law of induction. Because resistivity in the bulk superconductor is zero, the induced current produces the perfect diamagnet.

The necessity of the London penetration layer is seen in the requirement for organization of a current with a corresponding field along any direction in three-dimensional space. In the penetration layer, it would not be possible to organize an opposing field with the field axis, for example, contained within the layer and approximately parallel to one of the surfaces of the superconductor. This was first recognized by the London brothers.

Magnetic susceptibility in superconductors is at the heart of superconductivity, considering the fact that bulk superconductors are the only materials that have a volume magnetic susceptibility of –1. The connection between magnetic susceptibility and heat capacity for superconductors is more subtle. Superconductor heat capacity and magnetic susceptibility have identical critical temperatures. The facts that (1) superconductor heat capacity is higher than normal metal heat capacity, and (2) superconductor magnetic susceptibility occupies an extreme position (–1) on the scale of susceptibilities are both circumstantially related to the fact that superconductors have zero scattering of conducting electrons. We have not found any other factor, with documented changes at T_c, that could account for the change in magnetic susceptibility of the phase.

Relatedness of heat capacity, C_P, volume magnetic susceptibility, χ, and electron scattering from the conduction band can be seen from plots of heat capacity versus temperature for the second-order phase transitions to both superconductivity and ferromagnetism. Both of these transitions show local maxima in heat capacity at the respective phase transitions, though the details are specific to superconductivity or ferromagnetism. Fritz London pointed out the importance of the exchange interaction in superconductor phenomenology [4]. For superconductors, electron scattering is phenomenologically connected with the changes in magnetic susceptibility that occur at the phase transition. It seems that it will be essential to include communication of local magnetic properties within the Condon domain through the exchange interaction in a detailed theory of superconductivity.

Using the simple molecular orbital approach presented here, it has been possible to develop an approximate model for understanding the order of the critical temperatures of the one bar superconductors in the periodic table. Robust superconductors[*]

---

[*] Robust elemental superconductors have T_c between 9.25 and 1.5 K. The upper bound is the T_c of niobium, Nb, the highest value for the elements. The lower bound is an arbitrary choice.

at one bar are found in columns 3, 5, and 7, and columns 12, 13, and 14 of the periodic table. The two sets of columns are separated because each represents a single superconductor subgroup in the periodic table. At low temperatures, ~4 K, at least some elements in both of these groups have closed s subshells in the metal structure. Elements in columns 4, 6, and 8, the even electron columns in the first half of the d transition series, have open s subshells at low temperatures and are limited to being weak or very weak superconductors because of s-basis conducting wave function emergence in the active conduction band, which quenches superconductivity. The main group metals, columns 12, 13, and 14 in the periodic table,* start with zinc and cadmium, which are weak superconductors because of an open ns^1 subshell on the atoms of the conducting metal. At sufficiently low temperatures, conduction wave functions using these basis functions are not accessible; np-basis conduction bands are available and the elements become superconducting. Mercury in period 6 of column 12 has an unexpectedly high T_c, at 4.15 K. The high T_c for mercury points to something unusual in the mercury density of states. Mercury also has an unexpectedly high electrical resistivity, $96 \cdot 10^{-8}$ $\Omega \cdot$m.†

A maximum T_c for the elements of roughly 10 K is understandable in view of a simple treatment of partial wave scattering as a function of temperature and atomic basis functions.‡ All of the elemental superconductors appear to us to use conduction bands derived from p atomic basis functions. The highest T_c for a system that appears to utilize p-basis conduction bands is approximately 41 K for MgB_2.§

It is not surprising that controlling factors in atomic structure extend to the appearance of elements in the list of type II superconductors. Niobium, Nb, technetium, Tc, and vanadium, V, are all type II superconductors and middle transition metals.¶ Niobium and technetium have the highest and second highest superconducting critical temperatures at one bar for the elements. Vanadium has the fifth highest critical temperature for the elements. Its T_c is lower than those of lead, Pb, and lanthanum, La, in addition to niobium and technetium. Tantalum, Ta, shows some similarities to type II superconductors.

High-temperature superconductors in the cuprate class have shown d-wave properties for the superconducting wave function in numerous experiments since 1995 [5]. Both the temperature range and the d-wave characteristics of these superconductors are consistent with the view that the conduction band that carries the supercurrent is made up of copper (2+), Cu II, $3d$-basis functions that are empirically known to have zero partial wave scattering resistance at temperatures in the range of 140 K down. It is conceivable that f-basis very high-temperature superconductors could be produced from materials with tripositive lanthanide or actinide ions. Problems associated with weak coupling in the f-basis conduction band will be extreme in these materials. Experiments at elevated pressures designed to increase the f-basis conduction band coupling may be worthy of exploration.

---

* Column 10 elements are included in the "main group" because in the atomic state their electron configuration has a closed s and a closed d subshell, like the historical main group elements.
† See Table 8.1.
‡ See Table 8.2.
§ See Chapter 12, Figure 12.2.
¶ These three elements are in columns 5 and 7 in the periodic table.

Efforts have been made to produce a balanced presentation of the subject from the points of view of both physics and chemistry, the perspective at the outset is, however, undeniably chemical. A major effort has been made to balance this biased beginning by adoption of at least some of the language of physics in this area. For example, physics conventions on directions of current flow are used in the figures. It is our hope that this effort will form a foundation for more extensive collaboration between physicists and chemists in this important area of research.

You might wonder, what were the accidents of fate that kept the condensed matter community in both chemistry and physics from recognizing the importance of partial wave scattering to the development of electrical resistance in metals for 60 or 70 years? We have had a careful look at these questions as well. Electron scattering between Bloch, electron gas electrons, and phonons was introduced in 1928 [6,7] and has been in use continuously in the condensed matter community since. The initial report [8] that ultimately led to the development of the method of partial waves [2] was published 2 years later on the subject of atomic spectroscopy, not the properties of metals. Historically, the utilization of the method of partial waves has been strongest in the spectroscopy literature. Sommerfeld's major paper [1] presenting the quantum mechanical electronic foundations for the Wiedemann–Franz law was published in 1927. The focus of the manuscript and work that followed the Sommerfeld publication was on the establishment of quantum mechanics as the new paradigm. It was not so strongly focused on the thermalization of dissipative electron scattering as a subject per se. For that simple reason the significance of the Sommerfeld relationship* for our understanding of the nature of resistivity in metals has not been noticed until very recently.

By 1957, the electron gas–phonon paradigm was so deeply entrenched that there was no possibility of considering alternatives in discussing resistivity in metals. Electron gas wave functions do not have the full capacity to distinguish the differences between niobium and silver metals. Silver is the best conductor among the normal elemental metals. Niobium is the most robust elemental superconductor. The real differences are large. This is the origin of the theoretical problem for superconductivity. The historical developments in condensed matter were so strongly focused on the similarities of all metals and the electron gas model that little attention was given to the real differences that can be seen between different metallic elements. We have done our best to notice the efforts of those who investigated the differences. They laid the foundation for all current and future work.

Superconductivity is a challenging, stimulating, and very rewarding area for study. Although it is more than 100 years since the discovery of the effect by Heike Kamerlingh Onnes [9–11], at this point there is still much more work to be done in this field than has been published thus far.

Ralph Dougherty
J. Daniel Kimel
Tallahassee, Florida

---

* See Chapter 6.

REFERENCES

1. A. Sommerfeld, *Die Naturwissenschaften*, 1927, *15*, 825–32.
2. See, e.g., J. J. Sakurai, *Modern Quantum Mechanics*, revised ed., S.F. Tuan, Ed., 1994, Addison Wesley Longman, New York, 7.6 Method of Partial Waves, pp. 399–409.
3. W. Meissner, and R. Ochenfeld, *Naturwissenschaften*, 1933, *23*, 787–8.
4. F. London, *Phys. Rev.*, 1948, *74(5)*, 562–73.
5. D.A. Wollman, D. J. van Harlingen, J. Giapintzakis, and D. M. Ginsberg, *Phys. Rev. Letters*, 1995, *74(5)*, 797–800.
6. F. Bloch, *Z. Physik*, 1929, *52*, 555–601.
7. F. Bloch, *Z. Physik*, 1930, *59*, 208–14.
8. E. Fermi, *Nature*, 1930, *125*, 16–17.
9. H. Kamerlingh Onnes, *Leiden Comm.*, 1911, 120b, 122b, 124c.
10. P. F. Dahl, *Superconductivity*, 1992, American Institute of Physics, New York.
11. H. Rogalla, and P. H. Kes, Eds., *100 Years of Superconductivity*, 2011, Taylor & Francis, Boca Raton.

Acknowledgments

Many people have made significant contributions to this book for which Dan and I are both very grateful. Lance Wobus, our editor at Taylor & Francis, has provided clarity and direction in the execution of this effort. Without his skill this book would not have been completed. L. Will Evans provided the bulk of the new illustrations that appear here. Dr. Stephanie Dillon generated numerical data from graphical sources in the literature to support the analysis. Thomas Whitaker carefully read the manuscript and made clarifying suggestions.

Scientists and mathematicians who have served a consulting capacity on widely varying scales have significantly added to the depth of this effort. Those involved include Igor Alabugin, James Brooks, Eun Sang Choi, Ronald J. Clark, Naresh Dalal, Zachary Fisk, Robert L. Fulton, Ivar Giaever, Stephen Hill, Louis N. Howard, Philip Kim, Klaus von Klitzing, Harry Kroto, Eric Lochner, Alan Marchand, Stephen McGill, Stephan von Molnar, Xufeng Niu, Jihyung Shin, Gwo-Ching Wang, and Wang-Jie Wang.

The substantial support of the faculty, students, and staff of Florida State University and the National High Magnetic Field Laboratory is gratefully acknowledged.

1 Introduction

This book is about the theory of the superconducting state, and how to understand it within the context of the larger picture of physics, chemistry, and quantum mechanics. A superconductor passes an electric current with zero resistive heating of the conductor. This means an electric resistance and/or magnetoresistance of zero in the conductor. Under these conditions, movement of electrons in the conductor is nondissipative. The superconducting state is known to be massively entangled, which means that the Born–Oppenheimer separation of vibrational and electronic wave functions does not apply to superconductors.

Superconductivity was discovered in 1911 by a Dutch physicist, Heike Kamerlingh Onnes [1,2]. Superconductivity in solid mercury was discovered at ~4K in an experiment on the temperature dependence of electrical resistance. Kamerlingh Onnes' first major objective at the time was to extend knowledge in low-temperature physics to the lowest possible temperatures. The discovery centennial for superconductivity was celebrated in 2011. Articles and reviews appeared in major journals [3], and substantial, commemorative volumes were published [4]. These works are focused on the latest developments in the field like iron pnictide* and intermetallic† superconductors.

The theoretical paradigm for superconductivity has been in place for more than 50 years. The existence of unsolved, long-standing problems point to areas of potentially fruitful investigation. Problems like an effective theory of the superconducting critical temperature, T_c, or a theoretical understanding of the foundations for high-temperature superconductors, $T_c > 77$ K, have been open for many years. It is our hope that the molecular quantum mechanical approach suggested here will offer a theoretical basis that can lead to solutions of these and other problems in the field.

Modern superconductivity theory is referred to as BCS theory, after the initials of the author's last names, Bardeen, Cooper, and Schrieffer [5]. This complex theory posits the formation of "bound," spin-paired pairs of conducting electrons in the metal conduction band. The attractive force that binds the electron pair, a Cooper pair, is mediated by the nuclei in the lattice.‡ In principle, as temperature decreases, the number of Cooper pairs increases, and ultimately the superconductor becomes a fully spin-paired, diamagnetic system.

FIVE UNRESOLVED PROBLEMS IN SUPERCONDUCTIVITY

There are five as yet unresolved areas that we question regarding superconductivity, all involving a theoretical basis, or foundation in phenomenon:

---

* Iron pnictides are compounds of iron with members of the nitrogen family, N, P, As.
† Intermetallic compounds are metal alloys that have a definite chemical structure and formula, such as, $CeIrIn_5$.
‡ This is the same process that is involved in the formation of the widely studied covalent bond.

1. the pattern of elemental superconductors at one bar in the periodic table;
2. high-temperature superconductivity, superconductors with critical temperatures, $T_c > 77$ K;
3. unpaired electron spin in superconductivity, for example, the Knight shift in superconductors,[*] and the existence of triplet superconductors with nonzero electron spin;
4. higher specific heat capacity at constant pressure in superconductors as compared to the normal metals at T_c, coupled with magnetic susceptibility, $\chi = -1.0$,[†] for the superconducting state; and
5. superconductivity's alignment with the rest of atomic and molecular quantum mechanics, e.g., compatibility of superconductivity with the Sommerfeld equation and the quantum mechanical, electronic foundations of resistivity/conductance in ordinary circuits.[‡]

To a novice these five questions or declarations will not make much sense. Each statement is the subject of an introductory brief chapter that discusses the meaning of the statement. Chapters 2–6 show why the questions are problematic for the conventional view of superconductivity. They also point to a view that could lead to a deeper understanding. A desire to develop and expand this understanding is the reason for revisiting the subject of superconductivity, from its most fundamental foundations to current practice.

Our approach to problems in the realm of superconductivity is based on the chemical bonding theory. What is it that holds materials, including superconductors, together? This theory is atomic and molecular quantum mechanics. Atomic and molecular quantum mechanics is arguably the largest and most successful of modern scientific theories. The number of facts that are explicitly accounted for by this body of theory is very large. These facts include the structures and properties of the known organic and biological molecules, which at this time exceed 5×10^7 in number, a larger number than the number of documented species. The precision and accuracy of the theory are also beyond compare. Modern quantum mechanical descriptions of atomic and molecular structure and properties have been widely used in condensed matter physics since the early publications of Wigner, Slater, and Feynman, to mention only three of the early leaders in this field.

New insights into the nature of bonding in metals are presented and contrasted with bonding in insulators. Semiconductors fall into the continuum between metals and insulators; some are closer to metals and some closer to insulators. Quantum

---

[*] Knight shifts are shifts in nuclear magnetic resonance (NMR) frequencies for nuclei that are caused by unpaired electron spin.

[†] Magnetic susceptibility, $\chi = -1.0$, has never been seen in normal, fully electron spin-paired material, such as cellulose, benzene, or countless other closed shell fully electron spin-paired materials all of which are weakly diamagnetic. Many metals that are excellent conductors, including silver, are weakly diamagnetic, like spin-paired materials.

[‡] It will be essential for you to read at least the five following introductory chapters to develop an idea of what we are talking about concerning the quantum mechanics of resistivity in metals. The current model for electron scattering-caused resistance is based on the Bragg approach to scattering, which is classical; it does not require intimate quantum mechanical interaction of the electron with the atom from which it scatters.

mechanical understanding of magnetic susceptibility, within the structure of the phenomenological information that is available on the subject, is essential to understanding superconductivity. Thermodynamics of superconductivity is also centrally important. If we cannot explain why superconducting tin has a larger electronic heat capacity than normal metallic tin at nominally the same temperature and pressure, we are missing an essential part of the process (see Chapter 9). In the superconducting phase transition, a dissipative normal metal conductor is converted to a nondissipative superconductor. This process in the phase transition generates a temperature-dependent energy gap that is, at least in part, due to the component of free energy change resulting from differences in both enthalpy and entropy for the normal metal and the superconductor. The difference in entropy is the change in electronic entropy associated with the transition from a dissipative to a nondissipative current. The same statement is true for the enthalpy part of the free energy equation. The bulk of the energy gap is generated by the change in volume magnetic susceptibility from a nominal absolute value in the range of 10^{-5} to the diamagnetic extreme, -1. Magnetic susceptibility is discussed at length in Chapter 9. There is a section in Chapter 11 titled "Thermodynamic/Electronic Contributions to the Superconductor Energy Gap" that deals with models for the superconducting energy gap.

Metals are good conductors of unpaired electrons and are ductile and lustrous. These three physical properties are all linked to each other through the single electron bonds that hold metals together. To the extent that information concerning molecular bonding states can shed light on the processes in superconductivity, answers or directions for new research will appear for the five questions raised concerning the subject.

There is a deep principle of questions and answers in empirical science that can be stated approximately as, "When properly framed, the question is the answer." It is our hope that the five unresolved questions that form the heart of this effort will shed light on contested areas of superconductivity theory, so that the sources of the phenomenon will be clarified.

Our discussion begins with the first of the five unresolved areas about superconductivity that were raised above: the pattern of one bar superconductors in the periodic table of the elements. If you do not know the accepted mechanism of superconductivity in metals, do not be concerned. The essential concepts needed for understanding the discussion will be introduced with minimal mathematics.

This field involves only a few thousand active research scientists worldwide. A much larger numbers of people have been diagnosed by use of magnetic resonance imaging (MRI), in a superconducting magnet for medical data gathering. By reading this book it should be possible for you to have at least a beginning understanding of the theory of superconductivity as seen in a frame involving the full quantum mechanical behavior of conducting electrons in the system. The five unresolved questions present some of the problems that are available in the field for study. The following chapters construct a foundation for evaluating these problems, and provide a level of understanding of the form for solutions to the questions that have been raised.

One of the problems facing us is the real chasm between the language of physics and that of chemistry. Internationally, physics utilizes the sense of electric current flow that was proposed by Benjamin Franklin, that is, current flows in a conductor, from the positive pole to the negative pole. Historically, chemists abandoned that convention after the discovery of the electron by J. J. Thompson in 1897. As a consequence, in chemistry, current flows from the negative pole to the positive. This single primary difference cascades to different conventions for dipole moments, optical rotation, and other charge-carrier based conventions. A concerted attempt has been made to utilize the language of physics in such a way that the material is directly understandable to people with primary training in chemistry and physics, or more general training in science.

The first question that we raised is the pattern of elemental superconductors at one bar in the periodic table. Discussion of this pattern will be used to introduce the concepts of metals and insulators, in addition to conductance and resistivity. All of these concepts are essential prerequisites for understanding superconductivity. To develop these subjects, it will also be necessary to briefly review the quantum mechanical foundations of the periodic table of the elements. The historical origins of these observations, as a source of questions for superconductivity theory, stem from the fact that BCS theory uses electron gas wave functions that do not consider the atomic nucleus.*

REFERENCES

1. H. Kamerlingh Onnes, *Leiden Comm.*, 1911, 120b, 122b, 124c.
2. For an excellent historical account, see, P. F. Dahl, *Superconductivity,* 1992, American Institute of Physics, New York.
3. See, e.g., "Happy 100th Superconductivity," multiple authors, news and reviews, *Science*, 2011, *332*, 189–204.
4. See, e.g., H. Rogalla, P. H. Kes, Editors, *100 Years of Superconductivity,* 2011, Taylor & Francis, Boca Raton.
5. J. Bardeen, L. N. Cooper, J. R. Schrieffer, *Phys. Rev.*, 1957, *108*, 1175–1204.

---

* It would be surprising if it were possible for the BCS theory to account for the pattern of superconductors in the periodic table in view of the fact that it utilizes electron gas wave functions.

2 Pattern of Elemental Superconductors in the Periodic Table

Table 2.1 is a periodic table of the elements, with the elements that are superconductors at one atmosphere, and have critical temperatures, $T_c > 0.1$ K, shaded [1]. The lanthanides and actinides, elements 58–71, and 90–103, respectively, have a high density of electronic states compared to the rest of the periodic table. This feature, along with many unpaired electrons, make these elements particularly difficult to categorize theoretically, so the four one-atmosphere superconductors in the group have not been included in the following tables and will receive little attention here.

It is not difficult to come up with the idea that there is a pattern in the periodic table that is associated with superconductivity. Thus far, this pattern has not been theoretically accounted for in the literature. Existence of a pattern was pointed out many years ago by Barend Matthias (who was then at Bell Laboratories [2]), among other scientists who also worked on the problem. A connection between angular momentum and transition metal superconductivity was suggested by Hopfield in 1969 [3], using standard Bloch electron gas wave functions. The focus was on the number of valence electrons per atom, as pioneered by Matthias. Valence electrons are the electrons with the highest principal quantum number, n, for the element.

The first challenge for any attempt to theoretically explain the pattern of superconductivity in Table 2.1 is to clearly distinguish between the features that cause some elemental metals to be metals while others are not. Niobium, Nb, which is often included in lists of refractory metals, is the most robust (highest T_c) superconductor in the periodic table. Just six columns to the right you find silver, Ag, the best normal metal conductor in the table. Silver has never been shown to be a superconductor under any conditions. The fact that this distinction cannot be made with current theory must be seen as a limitation.

Analysis of the extremes in T_c shows that niobium, Nb, has the highest reported critical temperature for an element (9.25 K) at one bar. This maximum is related to the fact that Nb is in column 5 of the periodic table, the only column in the table that contains three robust superconductors ($T_c > 1.5$ K).

The minimum value for one bar T_c for bulk superconductors is held by beryllium, Be, $T_c = 26$ mK. Tungsten is not a bulk superconductor, but in a film, tungsten, W, has a T_c of 10 mK [1]. Theoretical predictions or rough estimates for critical temperatures for elemental superconductors are not presently available in the literature. See Chapter 8 for a discussion of the factors involved in the critical temperature for elemental superconductors.

TABLE 2.1

Periodic table showing bulk elements superconducting at one atmosphere, with their critical temperatures, T_c.

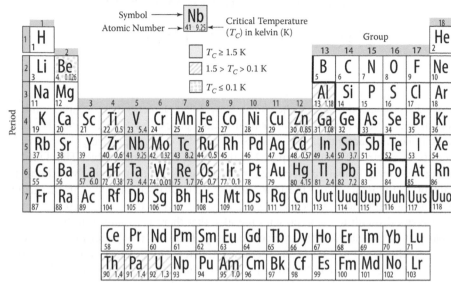

Note: Critical temperatures, T_c in kelvin, are shown at the bottom right for each element. Tungsten, W, is only superconducting in a film, not in bulk samples.

Source: C. Buzea and K. Robbie, *Supercond. Sci. Technol.*, 2005, *18*, R1–R8.

The pattern shown in Table 2.1, and the correlations reported by Matthias and Hopfield are not accidents. They are based on the precise electronic states of the conductors at the critical temperatures, T_c, for superconductivity. In order to appreciate this, it will be necessary to use wave functions that reflect the electronic features of the atoms or materials that they describe. In Table 2.1, the superconductors, the shaded elements, are all metals. They conduct electricity more or less effectively at room temperature. Univalent metals are missing from the category of superconductors.* The univalent metals include the members of the first column of the table, the alkali metals, and the eleventh column, the "noble" metals,† also called the copper family. Copper (Cu), silver (Ag), and gold (Au) have never been found to be superconductors, though they are the best ordinary conductors available. They have the lowest resistivity.‡ There are no one-bar superconductors in the divalent metals, the alkaline earths (column 2 of the periodic table), with the exception of the very

---

* Valence refers to the ability of an element to form chemical bonds. Under normal conditions, univalent elements form a single chemical bond, divalent elements form two bonds, and so on.
† Coinage metals is another term that has been used to describe the metals in column 11 of the table.
‡ Silver is the lowest resistivity elemental conductor, followed by copper and gold. The scientific literature presently lacks an accepted explanation of why these elements are not superconductors. Resistivity is a bulk property and has units, Ωm^{-1}.

weakly superconducting beryllium, Be. The divalent, zinc family (column 12 of the table) are all superconductors at some level. The highest T_c in the zinc family is that for mercury, 4.15 K. The cobalt and nickel families (columns 9 and 10 of Table 2.1) can have more than one valence, one of which is two. None of these elements are robust superconductors, $T_c > 1.5$ K, marked by the solid shading in Table 2.1.

Tri- and tetravalent metals are both well represented by superconductors (see the main group elements, columns 13 and 14). In the transition metals, Group 3, column 3 has one formal superconducting member, lanthanum, La, critical temperature 6 K. Like mercury, Hg, in the zinc family, lanthanum is the only robust superconductor in its periodic group. Members of Group 4 are all weakly superconducting. There is no data on superconductivity for rutherfordium, Rf, a synthetic element that is radioactive in all known isotopes.

Very weakly superconducting elements are shown in Table 2.1 with a shaded-dot background. These elements have T_c at or below 0.1 K. Temperatures this low are difficult to attain. For the weakest member of the group, beryllium, Be, the T_c is in the 30 millikelvin range. Critical temperature for tungsten, W, superconductivity in film form is only 10 mK. Since tungsten is not a bulk superconductor its inclusion as such in Table 2.1 is somewhat problematic. Discussion of this entire group of elements will be deferred until Chapter 11, because the discussion involves analysis of emergence of ns-basis conduction bands at low temperatures. We note on closing this section that robust superconductors can be found in columns 3, 5, 7, and 12 for the d transition series, and in columns 13 and 14 for the main group elements.

Matthias [2] and Hopfield [3] focused on counting electrons. We have expanded this approach by paying attention to the atomic basis functions for the lowest-temperature conduction bands for the superconducting and nonsuperconducting elements. Focusing on the periodic table brings in the orbital angular momentum quantum number, l, the foundation for the column structure of the periodic table. We have also considered the relative values of T_c for different elements. The utility of counting electrons is directly related to the electronic states of the systems. We want to look into the structure of the periodic table and the electronic states of the elements. To do this it will be necessary to refresh knowledge of the atomic quantum numbers, n, l, m, and the electron spin quantum number. Physicists refer to the electron spin quantum number as m_s; chemists speak of the same number as s.

If you are well versed in quantum mechanics, just skip the next section.

CHEMICAL INTRODUCTORY QUANTUM MECHANICS

In the classically based Bohr model, and since the four atomic quantum numbers, QNs, have been known as n, principal QN; l, orbital angular momentum QN; m, magnetic QN; and s, for chemists, electron spin QN, or m_s, for physicists. This is the z-component of electron spin angular momentum, where the coordinate z is determined by the placement of an external magnetic field.

The quantum mechanical revolution in physics and chemistry started about the time of Niels Bohr's semiclassical model of the hydrogen atom. Bohr's model was soon followed by highly developed wave mechanical models for the hydrogen atom, in which the probability of the electron's location at a given time was characterized by a wave function. A wave function is a mathematical function that satisfies a wave equation, like the equations written by Schrödinger or Dirac during the development of quantum mechanics. It isn't necessary to know the details of wave functions or wave equations to develop a working understanding of quantum mechanics. It isn't even necessary to explicitly think about waves as Heisenberg illustrated with his matrix mechanics approach to the problem. You do need to know that the motion of small subatomic particles like electrons is, in many respects, wave-like. A wave mechanically congruent approach is essential for its proper characterization.

Wave equations for the hydrogen atom have an exact solution because the problem has only two elementary particles, the relatively massive proton and the electron. The solutions correspond to total energy, total electron angular momentum, the z-component of electron angular momentum, and the z-component of the electron spin angular momentum. Quantum numbers arise in each of the previously mentioned solutions, which are referred to as eigenvalues.

Orbital is a word that is used interchangeably with wave function. This word was coined by Robert Mulliken. Often atomic orbitals, also called basis functions, are given letter designations that arose from their signatures in atomic electronic spectroscopy. The orbital designations are: *s*, *p*, *d*, and *f*. These letters stand for *sharp, principal, diffuse*, and *fine*, as observed in the structure of the electronic spectral bands. The four different bands and types of orbitals correspond, respectively, to orbital angular momentum quantum numbers, *l*, for the orbitals, *s*, *p*, *d*, and *f*, respectively, for 0, 1, 2, and 3.

PERIODIC TABLE QUANTUM STRUCTURE

Chemists are keenly aware that the orbital angular momentum quantum number *l* plays a major organizing role in the periodic table. Rows, periods, in the periodic table are determined by the principal quantum number *n*. For the outermost electron of the atom, *n*, there is an increase by 1 on each step down the right edge of the table. It tells you the principal quantum number for the outermost electron for all of the elements in the row. This *n* also tells you the total number of nodes in the wave function. For the highest energy electron of an element, the last electron to be added to the atom, *l*, tells you which section of the periodic table you are in; *l* varies from 0 to $(n - 1)$, in steps of 1. For the highest energy electron, if *l* = 0, it tells you that you are in the first two columns of the periodic table; if *l* = 2, this covers the next 10 columns of the table, the transition metals; and if *l* = 1, you are in the last six columns of the table. Elements in the first two columns and the last six columns are referred to as main group elements. Switching around the placement of *l* = 2 and *l* = 1 happens because the 3*d* subshell fills before 4*p* because of relative electron energy. Elements 58 to 71 and 90 to 103, the lanthanides and actinides, respectively, have nominal *l* = 3 for the highest energy electron, the last electron to be added in the stepwise construction of the atom. For elements 57 and 89 in column 3, the energies of the lowest vacant *d* and *f* wave functions depends upon the number of electrons present. Because of the very small energy gaps, the chemical and physical properties of lanthanum, element 57, and actinium, element 89, are more like their *f* electron relatives, than normal transition metals, and they are grouped with the *f* elements by the International Union of Pure and Applied Chemistry, IUPAC. Unfortunately, the IUPAC current version of the periodic table, which attaches La and Ac to f series elements, misses the role of La as a robust superconductor at one bar. Attached to the f elements, the robust superconductor La seems out of place. In its earlier placement in column 3 of the table, La looks like it is exactly where it belongs (see Table 2.1).

The value of l tells you the number of radial nodes in the wave function for the electron in the region of the nucleus. If such a node is present, it means that the electron has zero probability of being found at the same geometric coordinates as the nucleus if its position is described by a wave function with $l > 0$. l also tells you the exponent of the radial wave function for an atom as r approaches 0. Radial nodes that include the nucleus strongly influence the shape and compressibility of wave functions. Other things being equal, the larger the l quantum number, the more compressible the wave function for a corresponding pressure-caused increase in electronic energy.

Another feature of the periodic table pattern of one-atmosphere superconductivity is the critical temperature, T_c, for the one bar superconductors in the table. The critical temperature is the temperature of the phase transition in the solid-state that leads to superconductivity. There needs to be a foundation in theory for estimating, or evaluating, the critical temperature. When partial-wave scattering [4], involving the angular momentum quantum number, l, is used as the foundation for resistivity, avenues for calculation of critical temperature details present themselves (see the discussions in Chapters 8 and 11).

Robust superconductors ($T_c > 1.5$ K) are found at higher values of n in columns 3, 5, 7, 12, 13, and 14. This observation follows the chemical statement of the "inert pair" effect; as you proceed down the periods in a single column of the table, the s electrons become less and less likely to participate in chemical bond formation [5]. Lead with valence 2, Pb(II) has more stable compounds than Pb(IV), for example. This effect arises from the relative size of the gap in energy between the s and p electrons. In poly-electron atoms, s electrons are have a broader electron distribution and lower energy, and the p electrons are higher in energy with a narrower average range for the same period. Near the top of the periodic table, p electrons and s electrons both form strong bonds. As n increases, the energy gap between p and s electrons increases. The empirical result is a contraction of valence by two electrons, or a removal of s electrons from bonding. Separation of s-basis conduction bands from active conduction could account for the bulk of the robust superconductors having $n > 4$. Vanadium is the only exception, with $T_c = 5.4$ K. A theoretical basis for this observation is discussed in Chapter 11.

The column pattern of the robust superconductors in Table 2.1 is, for the first part of the d series: 3, 5, and 7. The fact that there is a one-column gap between these three numbers is significant. From the point of view of electron bookkeeping, elements in column 5 have one more formal electron pair than those in column 3. Why this might be important will not be very clear until we have a look at the probable electronic ground states of the metals as conductors.

Looking at Table 2.1, the thought appears that the key to understanding the details of superconductivity may reside in the information that is stored in the periodic table and crystal structures of the superconducting and nonsuperconducting metals. This observation is not at all new. The inside of the front cover for many of the current condensed matter physics books is a periodic table of the elements, often with unit cell structures, electronic configurations, and details of the solid state. The critical question that we need to address is the lowest temperature conduction band for the element that concerns us. Finding the answer to this question involves more than a quick look at the table, but an answer is available. We need to understand the distinction that can be made in the periodic table between metals and insulators.

METALS AND INSULATORS

Metals are materials that have surface-to-surface delocalized unpaired electrons and, among many other associated properties, they conduct electricity. Insulators in contrast do not conduct electricity and lack surface-to-surface electron delocalization. Metals form in the solid state because it is thermodynamically the most stable state of the system. In a molecular orbital description, metals are characterized by a very high density of delocalized electron states that span both the vacant and the fully occupied electron energy states of the system. In metals, the molecular analog of Hund's Rule for atoms applies; that is, the most stable states for energy-degenerate unpaired spins are those with magnetically aligned spins. When an energy gap forms between the occupied and vacant wave functions for a material,* surface-to-surface delocalization of electrons ceases, and the material becomes an insulator. Insulators cannot form delocalized electronic states from surface to surface, because of the energy gap between doubly occupied, bonding states and vacant nonbonding or antibonding states. Sir Nevill Mott predicted the existence of a phase transition between metallic materials and the corresponding insulators in 1949 [6]. Mott transitions depend on the development of an energy gap between the filled and vacant wave functions based on a change in temperature, pressure, magnetic field, or electric field. On one side of the transition the material is a metal, and on the other side the material is an insulator. The insulator has an energy gap between the bonding orbitals and the vacant orbitals. Energy gap formation is a consequence of a fundamental change in bonding mode for the lattice (see Chapter 7). Mott was awarded the Nobel Prize in 1977 [7].

Mott transitions clarify the model for bonding in metals. To rephrase the definition of metals, a metal is a material that exhibits surface-to-surface delocalization of unpaired electrons and lacks an energy gap between filled and vacant energy states for all states except superconductivity, which has a superconducting temperature-dependent energy gap.† If we understand what a metal is, it will be easier to understand the conduction bands in a metal. Conduction bands are wave functions that can be constructed from linear combinations of the atomic orbitals (LCAO), for the astronomically large number of atoms in a metal lattice. Because of the huge numbers involved we will not write out any wave functions. The wave functions themselves are beyond our needs. We need to know that they can be formed by use of symmetry

---

* For any process that does not involve the phase transition to superconductivity.

† In the Mott insulator state the energy gap is formally temperature independent, like the electronic excitation energies of molecules. There are, of course, cases in which the energy gap is not large. These cases may exhibit some form of temperature dependance of the gap not associated with superconductivity.

and that conduction bands are always closely spaced in energy. Conduction bands with nearly degenerate states are an essential feature of all metals. For formation of a metal conduction band: 1) in the ground state of the atom, the basis functions for the conduction band must be vacant or occupied by a single electron; and 2) the basis functions must be at the valence level for the metal. The valence level is determined by the maximum principle quantum number, n, for the element. This is the row number for the element in the periodic table counting down from hydrogen, H, as one. With these two requirements for metal conduction bands, we are in a position to talk about the highest energy conduction bands for the elements. The essential data that we need to construct the conduction band is the ground state electronic configurations of the isolated atom of the element. This material is available in the periodic table inside the front cover of most modern books on condensed matter physics. You can also find the information in chemistry texts and Table 2.2.

INTRODUCTION TO BONDING IN METALS AND NONMETALS

We will be using a variation on the Feynman band structure of metals [8]. The Feynman formalism uses normal space coordinates and is compatible with wave functions using either electron position or electron momentum, p_e (see Equation [2.1]) as canonical variables.

$$k = \frac{p_e}{\hbar} \tag{2.1}$$

Wave functions that form the conduction bands are composed of standard linear combinations of the atomic basis functions at the appropriate coordinates for the lattice. We utilize the principle of minimal atomic basis sets for molecular wave functions [9]. Photoelectron characterization of bonding in molecules shows that minimal atomic basis sets are used in chemical bonding. For example, methane, CH_4, is held together by 8 bonding electrons, 4 from carbon, and 1 each from the hydrogens. Photoelectron spectra show three bonds between the 4 hydrogen s wave functions and the three carbon p wave functions. The three bonds are degenerate; their electrons have the same energy. A fourth bond forms between the carbon $2s$ wave function and all four hydrogen $1s$ wave functions. The electrons in this bond are at a much lower energy, because of the relatively low energy of the carbon $2s$ wave function. The bonds form with just carbon s or p wave functions, not with mixed sets of carbon s and p wave functions. Many chemists expected that four bonds would be formed using carbon $2s^1 2p^3$ hybrid wave functions and hydrogen $1s^1$ wave functions. These directional bonds are not observed with photoelectron spectra. The bonding that is observed utilizes the simplest available basis set on each atom. In conduction bands, this means that the ground state conduction band for copper will have a single atomic basis function for each atom, for example, Cu $4s^1$. We can reliably expect that the ground state wave functions for copper metal will be formed from linear combinations of Cu $4s^1$ basis functions.

The atomic electron configurations for the elements that concern us in the periodic table are listed in Table 2.2. Only the valence shell, maximum principal quantum

TABLE 2.2
Ground State Electron Configuration Atomic Metals, Periods 3–6

f Elements and inner subshells not listed

Superconductors			Nonsuperconductors		
At#	Name	Configuration	At#	Name	Configuration
13	Al	$3s^23p^1$	11	Na	$3s^1$
22	Ti	$4s^2$	12	Mg	$3s^2$
23	V	$4s^2$	19	K	$4s^1$
30	Zn	$4s^2$	20	Ca	$4s^2$
31	Ga	$4s^24p^1$	21	Sc	$4s^2$
40	Zr	$5s^2$	24	Cr	$4s^1$
41	Nb	$5s^1$	25	Mn	$4s^2$
42	Mo	$5s^1$	26	Fe	$4s^2$
43	Tc	$5s^2$	27	Co	$4s^2$
44	Ru	$5s^1$	28	Ni	$4s^2$
48	Cd	$5s^2$	29	Cu	$4s^1$
49	In	$5s^25p^1$	32	Ge	$4s^24p^2$
50	Sn	$5s^25p^2$	37	Rb	$5s^1$
57	La	$6s^2$	38	Sr	$5s^2$
72	Hf	$6s^2$	39	Y	$5s^2$
73	Ta	$6s^2$	45	Rh	$5s^1$
74	W	$6s^2$	46	Pd	$5s^0$
75	Re	$6s^2$	47	Ag	$5s^1$
76	Os	$6s^2$	51	Sb	$5s^25p^3$
77	Ir	$6s^2$	55	Cs	$6s^1$
80	Hg	$6s^2$	56	Ba	$6s^2$
81	Tl	$6s^26p^1$	78	Pt	$6s^1$
82	Pb	$6s^26p^2$	79	Au	$6s^1$
			83	Bi	$6s^26p^3$

number electrons are listed. They are the only electrons that have sufficient atom-to-atom overlap to generate a conduction band or a "localized" bond. Data presented in Table 2.2 is widely available. It can be found on the inside front cover of several current condensed matter physics texts. It is also available from the *Handbook of Chemistry and Physics* [10] and from Web Elements [11]. The atomic radii for the metals, obtained from their x-ray crystal structures, vary smoothly across the periodic table with small discontinuities for formation of closed and half-closed subshell structures. There is no evidence available from crystallography that would support the notion of electron delocalization in metals involving electron participation from any electronic shell other than the valence shell of the metal.

Table 2.3 lists our proposed minimal basis functions for the Fermi level conduction bands of the listed elemental metals.

There are a number of details in Tables 2.2 and 2.3 that require experimental verification, notably the number of carriers in the superconductors at the critical

TABLE 2.3

Ground State Conduction Bands, One Bar Metals, Periods 4–6

f Elements and inner-shell bands not listed

Superconductors				Nonsuperconductors			
At#	Name	Z^*	Band basis	At#	Name	Z^*	Band basis
13	Al[12,‡]	1	$3p^1$	11	Na[13]	1	$3s^1$
22	Ti	1	$4p^1$	12	Mg[13]	2	$3s^13p^1$
23	V	1	$4p^2$	19	K[13]	1	$4s^1$
30	Zn	1	$4p^1$	20	Ca[13]	2	$4s^14p^1$
31	Ga[†]	1	$4p^x$	21	Sc	2	$4s^14p^1$
40	Zr	1	$5p^1$	24	Cr	1	$4s^1$
41	Nb[15]	1	$5p^2$	25	Mn[12]	<4	$4s^14p^1$
42	Mo	1	$5p^1$	26	Fe[13]	2	$4s^14p^1$
43	Tc	1	$5p^1$	27	Co[13]	2	$4s^14p^1$
44	Ru	1	$5p^1$	28	Ni[13]	2	$4s^14p^1$
48	Cd	1	$5p^1$	29	Cu[13]	1	$4s^1$
49	In	1	$5p^1$	32	Ge[†]	<1	$4s^x4p^y$
50	Sn[12,‡]	2	$5p^2$	37	Rb[13]	1	$5s^1$
57	La[16]	1	$5p^1$	38	Sr	2	$5s^15p^1$
72	Hf	1	$6p^x$	39	Y	<2	$5s^15p^1$
73	Ta	1	$6p^2$	45	Rh	1	$5s^1$
74	W	1	$6p^1$	46	Pd	<1	$5s^x$
75	Re	1	$6p^1$	47	Ag[13]	1	$5s^1$
76	Os	1	$6p^1$	51	Sb[12]	<2	$5s^x5p^y$
77	Ir	1	$6p^1$	55	Cs[13]	1	$6s^1$
80	Hg	1	$6p^1$	56	Ba[13]	2	$5s^15p^1$
81	Tl	1	$6p^1$	78	Pt	<1	$6s^1$
82	Pb[12,‡]	2	$6p^2$	79	Au[13]	1	$6s^1$
				83	Bi[12]	<2	$6s^x6p^y$

* "Z" is the estimated number of charge carriers per atom for T < 4 K or at T_c, whichever is lower.

† Indicates a semiconductor or semi-metal.

‡ The decrease in the number of carriers from the value at room temperature [12] is caused by low-temperature *s* sub-shell contraction, the "inert pair" effect.

temperature, T_c, and the number of carriers in the nonsuperconductors at liquid helium temperatures, 1 bar, 4.2 K. It is our understanding that the number of carriers per atom for the superconductors must be an integer.

Formation of Table 2.3 from the data in Table 2.2 was done by the following procedure. In order for a lattice to be a conductor, there must be unpaired, formally nonbonding electrons* available at the valence level. This means that the ground state

---

* This specifically refers to the electron pair model for bonding. Bonds in metals are formed with electrons in singly occupied conduction bands. In metal bonding interelectron magnetic interactions are very important, whereas those same interactions are negligible in electron pair bonds.

of, for example, titanium, Ti, atoms would not be conducting because the valence subshell is closed, $4s^2$, so a single electron must be promoted to a higher energy valence level basis function. The new conducting ground state involves $4s^1$ and $4p^1$ basis functions. Only the highest energy basis function, $4p^1$, is shown in Table 2.3 for the superconductors. It is assumed that the unfilled s-basis bands are sufficiently low in energy so as not to be present in the conduction bands for the superconductor. In the nonsuperconductors in Table 2.3, both the s-basis and p-basis bands are shown. The difference between titanium in the superconductor column and calcium, Ca, in the nonsuperconductor column of Table 2.2 is the size of the energy gap between the valence level s and p orbitals.[*] In the case of titanium, there is a sufficient energy difference between the valence level s and p orbitals that it is possible at low enough temperature to find $4p$ basis conduction bands without also finding $4s$-basis conduction bands at the same energy. Calcium, because it is closer to the left edge of the periodic table, has a smaller energy gap between the $4s$- and $4p$-basis functions, so at one bar it is not possible to find a temperature low enough to sample only the $4p$-basis functions at the top of the metallic conduction bands. In Table 2.3, this is indicated by the presence of both $4s^1$ and $4p^1$ conduction band basis functions for calcium, Ca.

Lack of s-basis functions at the Fermi surface[†] for superconductors opens an avenue for investigating electronic aspects of resistivity in metals. Fermi "contact" has been associated with electron occupation of s-basis functions since its discovery by Enrico Fermi in 1930. The observation that an electron and an atomic nucleus have a finite probability of occupying the same geometric coordinates in space-time was first made in atomic spectroscopy [17]. Fermi was investigating the electronic spectroscopy of atoms, specifically the hyperfine lines in the spectra of gas phase atoms [18] caused by magnetic coupling of electron spin with spin of the atomic nucleus. When an electron and a nucleus have a probability of having the same geometric coordinates, they can exchange both angular momentum and kinetic energy. Exchange of angular momentum gives rise to hyperfine lines in atomic spectra and nuclear spin–spin coupling in nuclear magnetic resonance spectra. Exchange of kinetic energy gives rise to equilibration of thermal energy between electrons and the metal lattice (see Chapter 6 and Chapter 11.) In Chapters 6 and 11, we will illustrate how s-basis Fermi functions facilitate thermal equilibration with the lattice of any electron energy that is outside the Boltzmann thermal distribution. This observation will contribute to the basis for understanding the pattern of one bar superconductors in the periodic table. It will also make it clear that silver, the best normal conductor of electricity in the periodic table, will never be a superconductor. The Fermi surface for pure silver has a $5s^1$ basis function. Fermi contact is the only experimentally verified quantum mechanical mode for contact interactions between electrons and atomic nuclei in molecules or metals that are unperturbed by external particles.

The pattern of the one-bar superconductors in the periodic table appears as the result of the quantum mechanical structures of the elements in their standard one-bar

---

[*] Ultimately, the basis functions that are seen in conduction bands depend upon the density of states for the metal at the temperature and pressure of the experiment. This data is not widely available at present.
[†] Fermi surface electrons are electrons in the highest energy occupied state of a metal at the lowest attainable temperature and one-bar pressure.

states at low temperatures. The structure of the atoms, which is directly available in the periodic table, and the interatomic interactions in the condensed phases that form the conductors or insulators appear to be responsible for the pattern that appears. The lowest energy conduction band of the solid controls the possibility of superconductivity for an element. If that band has s basis wave functions, the lowest temperature conduction band for the system will always be dissipative because of Fermi contact electron scattering. If the lowest energy conduction band for the metal has only wave functions with orbital angular momentum greater than zero, $l > 0$, at a sufficiently low temperature, the element can be a superconductor.

PRESSURE EFFECTS ON SUPERCONDUCTIVITY

Pressure strongly influences the critical temperature, T_c, of superconducting elements, and pressure effects have received wide attention with both elements and compounds [1]. Pressure effects on band structure of solids are known to be associated with the differential energy compressibility of s-, p-, d-, and f-wave functions [19]. McMahon's observations can be directly applied to superconductivity, if the model for superconductivity includes electron orbital angular momentum. This observation provides further support for the direct involvement of orbital angular momentum, atomic quantum number, l, in the phase transition from normal conduction to superconductivity. As elevated pressure rearranges the relative energies of the conduction bands associated with different values of l, isolated states with appropriate l values become available and the phase transition to superconductivity occurs. This has been studied explicitly in the case of the high pressure superconductivity of individual alkaline earths, elements in column 2 of the periodic table [20]. Since the dominant theory of superconductivity does not yet include electron orbital angular momentum, these important insights have remained on the periphery of the subject.

This brings us to the second unresolved question for superconductivity, the mechanism for high-temperature superconductivity, superconductors with $T_c > 77$ K, which is the subject of the next chapter.

REFERENCES

1. C. Buzea and K. Robbie, *Supercond. Sci. Technol.*, 2005, *18*, R1–R8.
2. (a) B. T. Matthias, *Phys. Rev.*, 1955, *97*, 74–6; (b) B. T. Matthias, *Prog. Low Temp. Phys.*, *II*, 1957, Chapter V, 138–50.
3. J. J. Hopfield, *Phys. Rev.*, 1969, *186(2)*, 443–51.
4. J. J. Sakurai, *Modern Quantum Mechanics*, revised ed., S.F. Tuan, Editor, 1994, Addison, Wesley, Longman, New York, pp. 399–410.
5. http://en.wikipedia.org/wiki/Inert_pair_effect
6. N. F. Mott, *Proc. Phys. Soc. (London) A*, 1949, *62*, 416–22.
7. http://en.wikipedia.org/wiki/Nevill_Francis_Mott
8. L. Solymar and D. Walsh, *Electrical Properties of Materials*, eighth edition, 2009, Oxford University Press, Oxford; p. 106.
9. M. J. S. Dewar and R. C. Dougherty, *PMO Theory of Organic Chemistry*, 1975, Plenum, New York; p. 27.

10. W. M. Haynes, Editor, *Handbook of Chemistry and Physics*, 92nd Edition, 2011, CRC Press, Boca Raton.
11. http://www.webelements.com/
12. No Marder; Table 17.1, p. 499. inc ref 14
13. Yes Marder; Table 17.1, p. 499. inc ref 14
14. (a) A. T. Burkov and M. V. Vedernikov, Thermoelectric properties of metallic materials, in *CRC Handbook of Thermoelectrics,* D. M. Rowe, Editor, 1955, pp. 387–399, CRC Press, Boca Raton; (b) H. Landolt, R. Börnstein, *Numerical Data and Functional Relationships in Science and Technology, II*, Springer, Berlin; (c) I. S. Grigoriev, E. Z. Meilkhov, Editors, *Handbook of Physical Quantities*, 1997, CRC Press, Boca Raton.
15. E. Fawcett, W. A. Reed, and R. R. Sodken, *Phys. Rev.*, 1967, *159*, 333–9.
16. C. Reale, *Appl. Phys. A*, 1973, *2*, 183–5.
17. E. Fermi, *Nature*, 1930, *125*, 16–7.
18. (a) E. Fermi, *Mem. Acad. D'Italia*, 1930, *1 (Fis.)*, 139–48; (b) E. Fermi and E. Segré, *Mem. Acad. D'Italia,*, 1933, *4 (Fis.)*, 131–58.
19. A. K. McMahon, *Physica*, 1986, *139B & 140B*, 31–41.
20. (a) J. Wittig, R. T. Matthias, *Phys. Rev. Letters*, 1969, *22*, 634–6; (b) K. Shimizu, K. Amayai, and N, Suzuki, *J. Phys. Soc. Japan*, 2005, *74*, 1345–57.

3 High-Temperature Superconductors, $T_c > 77$ K

It is significant that the materials that are high T_c superconductors are refractory insulators at room temperature. Cuprates are a class of high-temperature superconductors that contain a copper oxide layer. In the crystal structure, the copper oxide formula is CuO. Each copper 2^+ ion is associated in a square planar array with four oxygen ions, O^{2-}. You can see such a formal CuO array as a copper 2^+ ion associated with the four nearest neighbor oxygens, near the center of Figure 3.1. Charge balance in the crystal is always exact for the neutral material. It involves cations other than just Cu^{2+} in the lattice.

The conductor's current moves within the weakly coupled layer of copper II ions (Cu^{2+}), avoiding the presence of the oxygen ions. At room temperature, and with no doping,[*] cuprates are antiferromagnetic[†] insulators. Every atom of copper II has one unpaired spin. The outer electron configuration for the ground state dication, Cu^{2+}, is $3d^9$. In these structures, the unpaired electrons on the copper ions are antiferromagnetically[‡] coupled in the crystal, so a conduction band cannot be formed. Introduction of small amounts of unpaired electron dopants into the matrix produces lattice defects in the structure and provides a route for uncoupling the antiferromagnetism. The effect of antiferromagnetism on the ability of a system to form a conduction band is crystal-axis dependent. The conductivity effect is also dependent on the spatial orientation of the atoms that are conducting, and those that are antiferromagnetically coupled. It is not possible for an atom to be both antiferromagnetically coupled, and a conductor in the same geometric coordinate. From this starting point, addition of a proper dopant to the matrix can convert the insulating material to a superconductor that has a critical temperature higher than the boiling point of liquid nitrogen, $T_c > 77$ K.

Once the antiferromagnetic coupling in the copper planes is removed by doping, it is possible for the system to form a d conduction band. A reasonable example is BSCCO-2212, a commercial, high T_c superconductor, with approximate empirical formula, $Bi_2Sr_2CaCu_2O_{8+\delta}$ (see Figure 3.1). The amount of oxygen in the formula, $8+\delta$, depends on the level of doping, δ. Once a d conduction band has been formed in

---

[*] Doping refers to addition of trace materials that add conducting electrons or holes.
[†] Rather than having unpaired electron spins aligned in adjacent unit cells, as in a ferromagnet, the unpaired spins in an antiferromagnet, adopt spin orientations that cancel each other's magnetic moment.
[‡] Antiferromagnetic coupling is a nonbonding electronic interaction that blocks electron conduction and creates paired electron spins at an approximate distance of the lattice constant for the crystal.

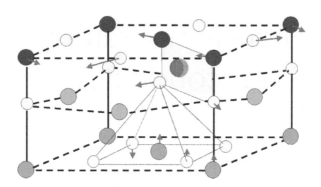

FIGURE 3.1 $Bi_2Sr_2CaCu_2O8+\delta$, partial crystal structure with a single interstitial oxygen (two-toned dark grey) which caused the atomic displacements shown. Atoms are: Bi (very dark grey[top layer]), stoichiometric O (off white), Cu (light grey [bottom layer]), and Sr (lightest grey [middle layer]). Arrows indicate rough measure of displacement of atoms from their positions without interstitial oxygen.[1] Complete unit cells for this have multiple copper planes and look like confused rainbows when color is used to identify the atoms.

the Cu II planes, it is a matter of lowering the temperature to T_c, the critical temperature, to observe superconductivity. Estimates of this temperature for different orbital types making up the conduction band can be seen in Chapters 8, and 11. The basis for these estimates is covered in Chapter 8.

d-WAVE SYMMETRY FOR HT_C SUPERCONDUCTORS

Prior to 1995, theory suggested that superconductors should show s band symmetry for the conduction band. In 1995 a group of physicists at the University of Illinois, Champaign–Urbana, probed the conduction wave symmetry in a Josephson junction constructed from superconducting YBCO, (yttrium barium copper oxide, $YBa_2Cu_3O_7$), another of the many high T_c cuprates [2]. The results of the experiment are shown in Figure 3.2. They prove that this high T_c superconductor has d wave symmetry. These results have been confirmed in detail many times in the past 15 years.

Both the critical temperature and the symmetry of the known high T_c superconductors can be given a semiquantitative understanding within the context of the theory of superconductivity presented here, (see Chapter 8 and Chapter 12). The central thesis of the theory is that resistivity in metals arises almost exclusively from the electronic wave functions of the system. These wave functions give rise to partial wave scattering, which is well known and documented. Partial wave scattering depends on electron orbital angular momentum, temperature, and pressure. The magnitude for partial wave electron scattering for an electron angular momentum state depends upon l, the absolute temperature, and pressure. States with s basis functions are the only states that are associated with Fermi contact, which will cause electron scattering at the lowest temperatures attainable. States with p, d, and f basis functions have no electron–nucleus interactions of this sort and show electron scattering that decreases into the noise at low temperatures, depending upon l. The amount of scattering for different basis sets at a given temperature follows the series $s>>p>>d>>f$.

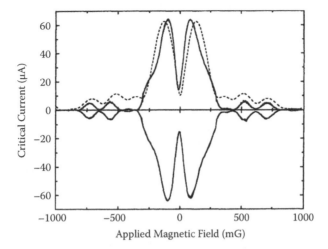

FIGURE 3.2   Current versus magnetic field for a corner junction, both polarities. Dashed curve calculated for $d_{x^2-y^2}$, assuming 15% asymmetry in junction area. (From D. A. Wollman et al., *Phys. Rev. Letters*, 1995, *74(5)*, 797–800.)

Because of these qualitative features of partial wave scattering, p conduction bands, which appear to be the foundation for superconductivity in mercury, have significantly lower critical temperatures than d conduction bands that are involved in the high T_c superconductors. It is informative that the d conduction bands of high T_c superconductors are reflected in the macroscopic symmetry of the electron waves in the conductor. See Chapters 8, 11, and 12 for more details.

An elevated critical temperature (T_c > 77 K) compared to elemental superconductors, and d-wave symmetry in the conduction band, are both natural consequences of adopting a modern molecular orbital approach to the study of resistivity and nondissipative electric conduction in superconductors. To be sure, consideration of such a model is a radical step for the current practice of condensed matter physics. The question is, do these suggestions make sense as straightforward answers to the questions concerning the T_c and the conduction band symmetry of the class of superconductors considered here?

Electron spin in a thin surface layer of a superconductor is responsible for what is known as the Knight shift. This phenomenon, a shift in the nuclear magnetic resonance frequency for a spinning nucleus caused by electron spin, was discovered during the course of Ph.D. thesis research by Walter D. Knight and has been widely studied for more than 60 years. Knight shifts and other electron spin effects in superconductivity are the subject of the next chapter.

REFERENCES

1. Y. He, T. S. Nunner, P. J. Hirschfeld, and H.-P. Cheng, *Phys. Rev. Letters*, 2006, *96*, 197002(1–4).
2. D. A. Wollman, D. J. VanHarlingen, J. Giapintzakis, and D. M. Ginsberg, *Phys. Rev. Letters*, 1995, *74(5)*, 797–800.

4 Electron Spin in Superconductors

UNPAIRED ELECTRON SPIN AND SUPERCONDUCTORS

Knight shifts are shifts in nuclear magnetic resonance frequencies of metal nuclei, caused by unpaired electrons in metal conduction bands. These shifts were discovered by Walter D. Knight in 1949 [1]. Knight shifts are obtained from magnetic resonance experiments on functioning superconductors. The Meissner–Ochsenfeld effect [2] prevents penetration of an external magnetic field into a bulk superconductor, so Knight shifts must be taken as surface measurements in thin films or powders, if the material is in a superconducting state. The electrons at the surface of a superconductor are not included in the macroscopic quantum mechanical entangled state known as superconductivity. If the electrons were a part of that state, it would not be possible to determine the magnetic resonance frequencies of adjacent nuclei and obtain the Knight shift because of the zero magnetic permeability.

There are two sources of the shift in magnetic resonance frequency for metal nuclei in conductors. The first has been referred to as a *contact* shift and is due to the presence of *s* basis orbitals in the conduction band for the nucleus under observation [3]. These paramagnetic shifts occur when electrons in *s* basis wave functions have a small probability of adopting the same geometric coordinates as the atomic nucleus. During the instant of contact between the nucleus and the electron, the two particles can, and do, exchange both angular (spin) momentum and kinetic energy. Exchange of angular momentum causes the Knight shift. Exchange of kinetic energy causes thermal equilibration of electron and nuclear kinetic energies.

The second cause of Knight shift is known as the *Pauli paramagnetic spin susceptibility* [3,4] and does not depend upon particle contact. This second component of the Knight shift is mediated through space and is not dependent (to first order) on the nature of the wave function associated with the electron. It is dependent upon the electron spin angular momentum, m_s or s.

The dominant theory of superconductivity holds that the Knight shift should be zero at absolute zero in superconductors [5]. The reason for this is that in BCS theory the elimination of resistivity in the metal is associated with forming spin-paired sets of electrons in the conduction band. At absolute zero, the electron-paired state should be the ground state, and electron pairing should be complete. Figure 4.1 illustrates the experimental Knight shift for superconducting tin, Sn, at temperatures below $T_c = 3.71$ K, as reported for the experiment [6].

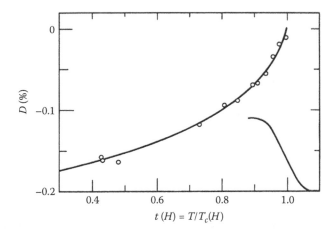

FIGURE 4.1   Knight shift, D (see Equation [4.1]) in superconducting tin.[6] The experimental points are for H = .012 T; $T_c(H)$ was 3.71 K. Solid line: a plot, of t(H) ~cosh[a(D − d)], a and d fitting constants. The lower curve is a plot of the change in frequency of an experimental oscillator (arbitrary units) as the temperature was lowered through T_c. Any correction in D for the demagnetization factor of the sample is less than 0.01%. (From the author's original paper.)

The Knight shift in Figure 4.1, is defined by Equation (4.1),[*] where ν_{sc} is the resonant frequency in the superconducting state, and ν_n is the normal state, 4.2 K, resonance frequency. The authors stated that "Any correction in D for the demagnetization factor of the sample was less than 0.01%."

$$D(\%) = \frac{\nu_{sc} - \nu_n}{\nu_g} 100 \tag{4.1}$$

Knight shifts are measured in superconductors in powder or film samples that have a high surface area compared to volume, because the Meissner–Ochsenfeld effect[†] excludes external magnetic fields from all but a thin surface layer of the metal in active superconductors. The experimental sample in Figure 4.1 was a film sample that contained alternating layers of vacuum evaporated tin and nylon. The tin layers were shown to contain small platelets of β-tin, that were ~140 Å in diameter and ~40 Å thick. The conducting allotrope of tin, β-tin, is a distorted face-centered cubic lattice; α-tin is a p-type semiconductor and is never a superconductor. The measured T_c for the sample was 3.71 ± 0.01 K [6].

Figure 4.1 shows a nonlinear drop in $D(\%)$ for values of $t(H)$ above 0.5, with an approximate linear decrease at lower temperatures.

---

[*] The tin sample was vapor deposited on nylon film. The resonance peaks in these experiments were very broad and featureless which is normal.

[†] See Chapter 9.

LOW-TEMPERATURE LIMIT OF KNIGHT SHIFT

In their 1961 paper on the low-temperature Knight shift in tin, Androes and Knight conclude that

> ...the electronic spin susceptibility in the superconducting particles at absolute zero is approximately three quarters of the normal value. The result for 1000 Å particles, though less accurate, is substantially the same [7].

The data in Figure 4.1 show that the Knight shift will not reach 0 shift at 0 K in the sample of tin examined. This means that there will be unpaired electrons available to the conduction band at the lowest attainable temperatures.

In the Knight shift experiment that we have been discussing it is probable that the entire tin sample was simply in the quantum Hall regime.* It is known that thin films of superconductors can support a state of zero electron scattering in the quantum Hall regime [8]. The energy gap formed in the quantum Hall regime has an identical origin to the corresponding energy gap in a superconductor, namely the change in electronic free energy associated with the change from dissipative to nondissipative current flow.† The tin sample used by Andros and Knight for generation of the data in Figure 4.1 was approximately 40 Å thick [6]. This dimension is almost certainly too small the sample to form a superconductor. Tinkham gives an estimate of the London field penetration depth for a typical sample as approximately 500 Å [9]. For the tin film samples used for the experiment of Figure 4.1 to become a superconductor, the sample would need top and bottom magnetic permeable layers. For the typical value of the London layer thickness of ~500 Å this would mean a 1000 Å-thick sample before you started to build the superconductor. Since the experimental film was reported to be ~40 Å thick there doesn't seem to be much hope of the sample becoming a superconductor, though it is very probable that electron scattering in the sample was zero at temperatures below 3.71 K. At temperatures below the superconducting critical temperature for thin film samples the quantum Hall regime is typically established, and there is zero electron scattering for conduction in the film.

The reflection above suggests that for tin metal, the conducting electrons in the quantum Hall state have unpaired spins, consistent with the Knight shift observations. Furthermore, the Knight shift measurements that have been conducted with thicker tin samples (in other reports), which gave similar results to those in Figure 4.1, would be consistent with superconducting tin having unpaired electron spins in the conduction band.

In considering the question of whether or not measurements of the Knight shift accurately reflect the spin state of the superconductor, we might carefully examine the one case that we are aware of, where spin determined by Knight shift and superconductor electron spin as determined using a Josephson junction have been compared.

---

* See Chapter 10.
† This subject is discussed in Chapter 11.

EXPERIMENTALLY VERIFIED ELECTRON
SPIN IN A SUPERCONDUCTOR

Directly related to the Knight shift in superconductors is the question of odd-parity superconductivity in strontium ruthenate, Sr_2RuO_4. Prior to the unequivocal verification that Sr_2RuO_4 is a triplet superconductor [10], the best evidence regarding the spin state of the superconductor came from studies of the Knight shift [11], which also showed it to be a triplet. The measurements made by Nelson et al. involved phase-sensitive use of a quantum interference device (SQUID) to establish with certainty that the superconductor had unpaired electron spin. Establishment of the triplet state of the current carrier in Sr_2RuO_4, by both a quantum interference experiment and the Knight shift, serves as support for the idea that the Knight shift can be used to report the spin state of the conduction bands in a superconductor. Recent observation of half-height magnetization steps in Sr_2RuO_4 [12], demonstrate the ability of the triplet spin of the superconductor to support half-quantum vortices in the structure. This nontrivial winding effect of the spin structure is another confirmation of the triplet spin state of superconducting Sr_2RuO_4.

Observation of unpaired spin in Sr_2RuO_4 as a superconductor are precisely coincident with the spin observations for the same material made using the Knight shift [11]. It is also true that prior to the direct measurement of spin effects in superconducting Sr_2RuO_4 by use of a Josephson junction [10], the question of triplet spin-pairing in the superconductor had been theoretically treated in the literature [11]. Nonetheless, the simplest explanation for all of the observations that have been reported to this time is that the spin state of the active superconductor and the spin state reported by the Knight shift in the London penetration layer are the same. Historically, the simplest explanation that accounts for all of the known facts is generally selected as the best explanation [13].

On the basis outlined above, it is our conclusion that the best evidence presently available indicates that superconductors have conduction bands made up of single electrons in molecular wave functions. The conclusion comes from evaluation of Knight shift data as well as data from experiments with Josephson junctions that indicate the Knight shift data provides an accurate report of the superconductor spin state. When the superconductor spin state is a triplet in Sr_2RuO_4, the Knight shift data from the London magnetically permeable layer, λ_L, reports a triplet. When the Knight shift for a superconductor reports unpaired spin, the superconductor just across the phase boundary has a high probability of having unpaired spin. In the case of the experiment reported in Figure 4.1, the data suggests that the measurements were actually made in the quantum Hall regime below the superconducting critical temperature so there was zero electron scattering in the sample.

Heat capacity for superconductors at the critical temperature, T_c, is known to be larger than the heat capacity of the normal conductor at a marginally higher temperature. The magnitude of the difference in heat capacity for superconductors scales with their critical temperatures. This observation does not yet have a clear explanation, nor does the observation of −1 magnetic susceptibility of bulk superconductors. These questions and the circumstantial relationship between them are introduced in the next chapter.

REFERENCES

1. W. D. Knight, *Phys. Rev.*, 1949, *76*, 1259–1260.
2. W. Meissner, and R. Ochsenfeld, *Naturwissenschaften*, 1933, *23*, 787–788.
3. See, e.g., E. Bekaert, F. Robert, P. E. Lippens, and M. Ménétrier, *J. Phys. Chem. C*, 2010, *114*, 6749–6784. This paper from *CERN* discusses the Knight shift in terms of Pauli-type susceptibility and Li *s* orbital participation in the density of states at the Fermi level leading to a Knight shift, *K*, which they computed using (see p. 6752, equation 1):

$$K = \frac{8\pi}{3} \left\langle \left| \Psi_{r=0} \right|^2 \right\rangle \chi_s^e$$

$\left\langle \left| \Psi_{r=0} \right|^2 \right\rangle$ is only non zero for the density of *s* basis electrons at the nucleus. χ_s^e is the Pauli susceptibility. The values were averaged over the Fermi surface.
4. W. Pauli, *Z. Phisik,* 1927, *43*, 601–23.
5. J. R. Schrieffer, *Theory of Superconductivity, Revised Ed.*, 1999, Perseus Books, p. 244–8, Section 8-8 The Knight Shift.
6. G. M. Androes, and W. D. Knight, *Phys. Rev. Letters,* 1959, *2(9)*, 386–387.
7. Quotation from the abstract: G. M. Androes, and W. D. Knight, *Phys. Rev.,* 1961, *121*, 779–787.
8. A. T. Bollinger, G. Dubuis,, J. Yoon, D. Pavuna, J. Misewich, and I. Božović, *Nature*, 2011, *472*, 458–460.
9. M. Tinkham, *Introduction to Superconductivity*, second edition, 1996, Dover, New York; p. 2, footnote 4.
10. K. D. Nelson, Z. Q. Mao, Y. Maeno, and Y. Liu, *Science*, 2004, *306*, 1151–1154.
11. A. P. Mackenzie, and Y. Maeno, *Rev. Mod. Phys.,* 2003, *75*, 657–712.
12. J. Jang, D. G. Ferguson, V. Vakaryuk, R. Budakian, S. B. Chung, P. M. Goldbart, and Y. Maeno, *Science*, 2011, *331*, 186–188.
13. R. Hoffmann, V. I. Minkin, and B. K. Carpenter, *HYLE—Int. J. Phil. Chem.*, 1997, *3*, 3–28.

5 Heat Capacity and Magnetic Susceptibility in Superconductors

The underlying subject of this introductory chapter is the mutual relationship between experimentally observed heat capacity and magnetic susceptibility for superconductors, and the thermodynamic consequences of dissipative or nondissipative conduction of electricity by metals. Normal metals are dissipative conductors of electric current. Superconductors are nondissipative conductors of similar currents. The transition from normal conductor to superconductor involves an explicit change in entropy for the conductor. The recent experimental demonstration of Landauer's principle, which links thermodynamics and information, by Lutz et al. [1] confirms the need for a change in system conduction free energy at the phase transition between normal metal and superconductor. We submit that this is one of the central questions that must be addressed in superconductivity. The three subjects introduced here are thus: free energy differences between dissipative and nondissipative conductors; the increased heat capacity of superconductors at the critical temperature, T_c, as compared to the precursor normal metal; and the extreme value of superconductor magnetic susceptibility, -1.0 (no units).

Free energy differences between dissipative and nondissipative current flow in metals do not appear to have been discussed in the literature thus far. Entropy differences between dissipative and nondissipative electron currents must contribute to the temperature-dependent energy gap that characterizes the nondissipative currents in superconductivity. This observation from the thermodynamics of superconductivity is intimately associated with the discussion of both the heat capacity of superconductors at T_c and the changes in magnetic susceptibility associated with the transition to superconductivity.

HEAT CAPACITY

Graphical results from an early report of heat capacity for the metallic phase of tin at the transition to superconductivity are presented in Figure 5.1 [2]. These results, reported by Keesom and Kok [2], are one of many examples illustrating the change in heat capacity at constant pressure, C_p, for a superconducting element at the critical temperature. For the superconducting transition in tin, see also Figure 9.8, which presents a 1956 data set on the same subject.

Increasing the heat capacity of tin by decreasing the temperature of the sample at the critical temperature for a superconductor is curious. What is it about the

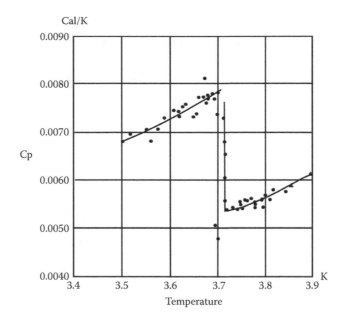

FIGURE 5.1   Heat capacity of tin as a function of temperature near T_c, = 3.72 K. (From W. H. Keesom, and J. A. Kok, *Akademie der Wetenschappen, Amsterdam, Proceedings*, 1932, *35*, 743–748.)

superconducting state that would cause a substantial increase in the heat capacity for a small decrease in temperature, in the range for tin, of 3.72 K, T_c? Heat capacity generally increases with increasing temperature, and the increases are usually smooth. In this case the increase is discontinuous and in the reverse sense from normal. It is worth recalling that a discontinuous change in heat capacity is an indicator of a phase transition.

Data in Figure 5.1 tell us that the superconductor is able to store more thermal energy per mol•kelvin, than the same number of particles in the same metal at a very slightly higher temperature. The difference between the heat capacities for the two states in Figure 5.1 divided by the maximum heat capacity of the superconducting state is 0.31. This means that 31% of the maximum heat capacity in the superconducting state goes missing in the phase transition to the normal state at T_c. In the superconducting state, there is no scattering of the conducting electrons. It is commonly known that electrons are the primary carriers of heat capacity at low temperature. Although separation of electronic and lattice heat capacity is not strictly valid for a superconductor, Figure 5.1 suggests that the scattered electrons in the normal metal are not carriers of electronic heat capacity.

What we are looking for here is a reasonable explanation for why the superconductor at T_c always has a heat capacity that is greater than the heat capacity of the normal conductor. The available data shows that the magnitude of the difference between the superconducting and normal heat capacities scales with temperature. If T_c is in the low millikelvin range, it may not be possible to measure the difference in heat capacity between the normal metal and the superconductor.

MAGNETIC SUSCEPTIBILITY OF SUPERCONDUCTORS

Observation of magnetic susceptibility, χ, of -1,[*] magnetic permeability of zero, in tin and lead cylinders, by Meissner and Ochsenfeld set the central defining observation for superconductivity [3]. Clarity of the theoretical foundation describing the quantum mechanical origins of magnetic susceptibility of -1 is hardly comparable. Although the -1 magnetic susceptibility of the superconducting state is arguably the most profound puzzle associated with superconductivity, the word *susceptibility* is mentioned only once in the first full paper launching the current theory of the subject [4]. The two earlier papers by the same authors do not mention susceptibility [5,6].

Superconductors are the only known materials with magnetic susceptibility of -1. Because superconductors exclude external magnetic fields ($\chi = -1$), it is not possible to directly observe the spin state of electrons in a superconductor. It is possible to obtain spin-free, fully bonded organic or inorganic chemicals that have fully paired electron spins. All of these materials that we have found data for have magnetic susceptibility with absolute value less than 0.01. Since virtually all simple spin-paired materials are diamagnetic, the imagined connection between spin pairing and the "pure diamagnetic" susceptibility of superconductors is a real part of the attractiveness of a spin-pairing theory.

A change in magnetic susceptibility, χ, from diamagnetic, $0 > \chi \gg -1$, or paramagnetic, $0 < \chi \ll 1$, to what has been called "pure diamagnetic," $\chi = -1$, is what the Meissner–Ochsenfeld effect is about [3]. If it involved only the pairing of electron spins, there are millions of known electron spin-paired compounds and materials that would have given us some indication of the validity of a spin-pairing hypothesis by this point in time.

Magnetic properties of materials are often difficult to understand, so they tend to occupy the bottom shelf. As a brief introduction, we will point to some of the data that is presently available. Molar susceptibilities, χ_M, for spin-free, fully electron spin-paired materials are typically in the range of -10^{-4} to -10^{-6} cm^3/mol (cgs units). Reliable values for elements and inorganic compounds are available from the Fermi Lab website [7]. Electron spin-free materials (fully spin-paired) are diamagnetic, but only weakly so. Volume magnetic susceptibility for a superconductor is -1. If you use the Fermi Lab list, remember that it contains materials, including elements, that are ferromagnetic and/or have unpaired electron spins. You should also recall that conversion of the values in the table to SI units involves molar volume conversion from cgs volume to SI volume. Conversion of cgs volume susceptibility to SI also requires multiplication by 4π.

Diamagnetism of zero electron spin materials is simple electron diamagnetism. It is a consequence of the fact that electrons have a negative charge. When a negative charge moves across a magnetic field the moving charge experiences a centripetal acceleration, which then generates motion of the charge to produce a magnetic field that opposes the externally applied magnetic field. Virtually all of the electrons in a zero electron spin material are subject to these effects and their motion combines to produce the observed diamagnetic response of these materials.

---

[*] Magnetic susceptibility (volume) is defined by Equation (5.1), and is unitless in SI.

There is general agreement that electron motion and spin in materials are the key to control of both magnetic permeability, μ, and volume susceptibility, χ, which are related to each other and to magnetization, M, as a function of applied magnetic field strength, H.

$$\chi_v = \frac{\delta M}{\delta H} = \frac{\mu}{\mu_0} = -1 \tag{5.1}$$

Magnetic susceptibility is, in general, a tensor, though most susceptibility reports give values as scalars. The second equality in Equation (5.1) is valid only when the magnetic response of a material is linear (no ferromagnets or nonlinear magnetic components). Magnetization in metals involves electron spin alignment in the lattice. In ferromagnets, electron spins in adjacent unit cells are magnetically aligned in magnetic domains.

Paramagnets and diamagnets are the materials that concern us as the foundations for superconductors. Paramagnets are materials in which electron spins increase their local alignment in parallel with an applied magnetic field. Diamagnetic materials increase their local alignment in opposition to and/or increase their diamagnetic electron orbital motion in response to an applied magnetic field. Elemental metals that are superconductors at one bar, and those that are not, are included in both diamagnetic and paramagnetic metal types. Experimental evidence relevant to the change in magnetic susceptibility for a normal metal to the bulk susceptibility of a superconductor, -1, is discussed in Chapter 9.

Magnetic domains in ordinary metals, specifically diamagnetic domains in metallic beryllium, were defined by J. H. Condon in 1966 [8]. Since that time, magnetic domains in diamagnetic and paramagnetic conducting materials have often been referred to as Condon domains.

Diamagnetism in graphite is known to be associated with diamagnetic currents in the planar graphite π-electron system [9].* Graphite is an archetype π-electron conductor. Conduction electrons in graphite move in π-symmetry wave functions that are nominally planar. Most metal conductors have σ-orbital conduction bands.[†] These wave functions are radially symmetric around the bond axis, and the σ wave function networks are typically three dimensional in solids. In the case of graphite, the diamagnetism is electron orbital diamagnetism, directly analogous to diamagnetic ring currents in the NMR spectra, of aromatic organic molecules—molecules like benzene[‡] that can be thought of as a small bit of monolayer graphene. Graphite diamagnetism is also directly analogous to Landau diamagnetism in an electron gas [10], see Chapter 10.

The fundamental question is, what is the physical origin of magnetic susceptibility in conducting materials? The answer is electronic communication of local

---

* The definition of a π-orbital from the glossary reads: A molecular wave function for a bonding or nonbonding electron that has a nodal symmetry plane that connects adjacent bonded atoms that are the source of the basis functions for the orbital. The π-orbital that lies above and below the plane of monolayer graphene is an archetypical example in condensed matter physics and chemistry.

† σ-orbitals have radial symmetry, to first order, between the bonded atoms.

‡ Benzene is C_6H_6, for a line drawing valence bond structure see Figure 8.3.

response to external magnetic fields throughout the Condon magnetic domain of the metal [8]. Electronic exchange interactions between electrons are the fundamental mechanism for electronic communication of electron spin information, which provides one basis for magnetic properties. Orbital angular momentum is a second and more subtle basis for magnetic properties of materials. Relationships that control the total angular momentum of domains of electrons in condensed materials are responsible for generating magnetic properties like magnetic susceptibility.

The only systems that exhibit zero electron spin correlation are bulk superconductors. These systems are not influenced by external magnetic fields below the critical magnetic field, H_c, and have magnetic susceptibility, $\chi = -1$. There is no requirement for electron spin pairing in solids that have magnetic susceptibility less than zero. Diamagnets include excellent conductors such as copper (Cu), silver (Ag), and gold (Au). Fully bonded organic molecules like cellulose are also diamagnets.[*] Metals have unpaired spins in their conduction bands. Spin-paired molecules can be prepared with purities high enough that no spins can be detected in the material. Both groups of materials have similar ranges of diamagnetic susceptibilities [7].

Material in Chapter 9 suggests that electronic angular momentum communication within the Condon domains of normal metals and the presence or absence of partially filled inner electron shells, d or f subshells, are responsible for segregation of elemental metals into diamagnetic and paramagnetic types. For superconducting metallic elements at T_c, communication of electron spin and electronic angular momentum response to external magnetic fields below H_c ceases and the magnetic susceptibility of both diamagnetic and paramagnetic metals becomes that of a bulk superconductor, $\chi = -1$.

There are subtle details of the quantum mechanical electronic foundations for superconductivity that are not found in standard textbooks. Some of these details are introduced in the following chapter.

REFERENCES

1. A. Bérut, A. Arakelyan, A. Petrosyan, S. Ciliberto, R. Dillenschneider, and E. Lutz, *Nature*, 2012, *483*, 187–189.
2. W. H. Keesom, and J. A. Kok, *Akademie der Wetenschappen, Amsterdam, Proceedings*, 1932, *35*, 743–748.
3. W. Meissner, and R. Ochsenfeld, *Naturwissenschaften*, 1933, *23*, 787–788.
4. J. Bardeen, L. N. Cooper, and J. R. Schrieffer, *Phys. Rev.*, 1957, *108*, 1175–204; the word susceptibility appears on p. 1189, col. 2, second full paragraph, in a discussion of the paramagnetic susceptibility in the Knight shift.
5. J. Bardeen, L. N. Cooper, and J. R. Schrieffer, *Phys. Rev.*, 1957, *106*, 162–164.
6. L. N. Cooper, *Phys. Rev.*, 1956, *104*, 189–190.
7. http://www-d0.fnal.gov/hardware/cal/lvps_info/engineering/elementmagn.pdf
8. J. H. Condon, *Phys. Rev.*, 1966, *145*, 526–535.
9. M. P. Sharma, L. G. Johnson, and J. W. McClure, *Phys. Rev. B*, 1974, *9*, 2467–75.
10. L. D. Landau, *Z. Phys. A*, 1930, *64*, 629–637.

---

[*] Condon did not consider molecules like cellulose [8]. A magnetic domain definition for these materials would not exceed the limits of the molecule and could be limited by electron mobility in, for example, high molecular mass insulators.

6 Quantum Foundations of Molecular Electricity and Magnetism

Courses on electricity and magnetism are generally available in physics and engineering departments. Molecular electricity and magnetism has recently been the subject of numerous special topics courses in both physics and chemistry where the study of molecular magnets like, Mn_{12}, has created substantial interest. We are interested in opening the discussion of molecular electricity and magnetism in the specific area of superconductors and related electronic states. The massively quantum entangled state of a superconductor represents just one molecule. Smaller pieces, such as the quantum Hall regime, are somewhat less complex but still out of the ordinary. The quantum mechanical requirements for getting in and out of these states are quite subtle and merit individual attention. Sommerfeld's relationship, which explicitly connects electrical conductance and thermal conductance, was the first part of this puzzle. It is still centrally important.

THE SOMMERFELD RELATIONSHIP

Sommerfeld derived the relationship between electrical conductivity, σ, and thermal conductivity, κ, at a given temperature, T (Equation [6.1]) [1]. Sommerfeld's relationship was based on the Wiedemann–Franz law. It was very important in gaining the acceptance of the quantum mechanical description of matter in the early days of the subject [1].

$$\frac{\kappa}{\sigma T} = \frac{\pi^2}{3}\left(\frac{k_B}{e}\right)^2$$

(6.1)

Thermal conductivity, κ, is related to electrical conductivity, σ, and absolute temperature, T, by the physical constants in Equation (6.1); π is the mathematical constant, k_B is the Boltzmann constant, and e is the charge on the electron. In his derivation of this relationship, Sommerfeld utilized pure electronic wave functions for an ideal electron gas. He made no use of any vibrational wave functions, or lattice phonons, the quantum mechanical carriers of vibrational energy in metals. Transverse phonons in solids are directly analogous to vibrational quanta in molecules. The question that Sommerfeld added to the mystery of the Wiedemann–Franz law is: "How are thermal equilibration of energy and resistance to electrical conduction accomplished using only electronic wave functions?" It is essential to understand the answer to this question if we are to fully understand resistivity in metals. The Wiedemann–Franz law is the historical precedent for the Sommerfeld relationship.

WIEDEMANN–FRANZ LAW

Wiedemann and Franz discovered in 1853 that metallic thermal conductivity divided by electrical conductivity has approximately the same value for different metals at the same temperature [2]. Twenty years later, 1872, Ludvig Lorenz discovered the proportionality between the ratio of thermal to electrical conductivity and temperature [3]. The proportionality constant is approximately 2.44×10^{-8} Watt·ohm per kelvin squared.

$$L = \frac{\pi^2}{3}\left(\frac{k_B}{e}\right)^2 = 2.44 \times 10^{-8} W\Omega K^{-2} \tag{6.2}$$

Equation (6.2) gives the value that one would obtain from the Sommerfeld relationship, which is usually called the Lorenz number. For real metals the values vary with temperature. Occupancy of specific conduction bands depends explicitly on the temperature. Only at the lowest temperatures is it possible for the active conduction band to be at the Fermi level.* The value of the Lorenz number also varies somewhat from metal to metal. Basis wave functions† in the conduction band at any given temperature depend upon the element, and the value of the physical constant depends upon the basis functions involved in the conduction band; that is, the density of states derived selection of s, p, d, and/or f atomic basis orbitals.

For copper, silver, and gold in the low-temperature range (2 to 30 K), the respective values of the Lorenz number are 2.45, 2.45, and 2.47×10^{-8} WΩK^{-2}. Values are from a review by Kumar et al. [4]. At higher temperatures, the deviations from the calculated value for the Lorenz number, Equation 6.2, become significantly larger. The Lorenz number also depends on pressure, as you might have expected from the fact that it depends on the conduction band wave mechanical basis set. The Sommerfeld relationship suggests that the equilibration of both thermal and electrical energy with a conducting lattice is primarily an electronic process that is governed by the electronic wave functions of the system.

Sommerfeld was establishing a relationship between electrical conductivity and thermal conductivity. Electrical conductivity involves electrical resistivity and thermal conductivity does not. The Wiedemann–Franz law shows that resistivity and thermal conductivity are related. In the case of thermal conductivity, there is no electric field and no source of electron acceleration other than the presence of atomic nuclei. Nonetheless, electronic quantum mechanics is the major control for thermal conductivity as well. Sommerfeld was working prior to the discovery of the quantum mechanical description of the electronic quantum scattering process, namely, the

---

* Fermi-level electrons are electrons in the highest energy occupied state of a metal at temperature approaching 0 K and one-bar pressure.

† In molecular quantum mechanics, the basis functions are the atomic orbitals that are used to create the linear combination of atomic orbital (LCAO) molecular orbitals, MOs.

method of partial waves [5]. This process has the capacity for a detailed explanation of the Sommerfeld equation in terms of the electronic quantum mechanics of metals.

Reality of the connection between the specific heat of metals and metal electronic properties including resistivity, magnetic susceptibility, magnetization, and superconductivity, is ultimately connected to the Wiedemann–Franz law by quantum mechanics. Sommerfeld was one of the first to seriously explore this connection. There is also little doubt that at the time of his formulation of the relationship shown in Equation (6.1), Sommerfeld was not clearly aware of the full implications of his work, particularly regarding the electronic aspects of heat transfer and thermal dissipation in metals.

ORIGINS OF RESISTIVITY IN METALS

There are four areas in condensed matter physics that highlight the foundations of resistive electron scattering in metals. These areas are: (1) the historical development of the subject; (2) the quantum mechanics of resistivity in metals, specifically resistivity that arises from electronic quantum mechanics, for example, the von Klitzing constant;[*] (3) existence of other electromagnetic quanta, related only to the electromagnetic properties of atoms in lattices; and (4) the concept of electron "effective mass" in experiments with metals involving external magnetic fields. These subjects will be briefly introduced in order.

HISTORICAL ORIGINS OF ELECTRON SCATTERING RESISTIVITY

Drude likened resistivity in electricity and magnetism to friction in mechanics [6a]. It was not long before Bloch introduced the idea of electron scattering, using a classical Bragg model, to describe the scattering interactions between lattice phonons and conduction band electrons [6b].

In Bloch's account of his thesis work on metallic resistivity some 52 years after the fact, he points out his awareness of the Sommerfeld relationship. Here, he is discussing an interaction with his mentor, Heisenberg:

> There was more of a challenge in another suggestion of his: to look into the theory of metals. During my earlier studies I had read the classical book of H. A. Lorentz on the theory of electrons and it was obvious that his work, based on Boltzmann statistics, had to be modified. Pauli had already shown that the application of Fermi statistics led to the temperature-independent paramagnetism of conduction electrons, but the most important applications were made by Sommerfeld. He had solved an old puzzle by demonstrating the great reduction from the classical value in the specific heat of a degenerate Fermi gas and, further, had developed the new consequences for the ratio of the electric and thermal conductivity of metals. Except for the replacement of classical statistics and the inclusion of the spin, however, Pauli and Sommerfeld both accepted the old ideas of Drude and Lorentz, who treated the conduction electrons as an ideal gas of free particles. The high conductivity and reflectivity of metals of course strongly supported the assumption of very mobile electrons but I had never understood how anything like free motion could

---

[*] See also, Chapter 10.

be even approximately true. After all, a metal wire with all its densely packed ions is far from being a hollow tube and as I started to think about it, I felt that the first thing to be done in my thesis was to face this striking paradox [7].

The quote above illustrates Bloch's depth as a scholar. It also illustrates that he was not aware of the implications of the Sommerfeld relationship and the Wiedemann–Franz law for the mechanism of electron scattering in metals. That Sommerfeld's electron gas model gave a good account of the relationship between electrical conductivity and thermal conductivity implies that for the bulk of heat transfer and electron scattering, neither one of the processes has a intimate involvement with the atomic lattice of the metal. If phonon involvement was a major requirement for either electrical or thermal conductance, it would have been essential for phonons to appear in some form in the derivation of the Sommerfeld relationship. This was not the case. This observation leads to the conclusion that the relationship memorialized by the Wiedemann–Franz law involves only the electronic (electron gas) structure of the metal.

Data cited throughout this book indicates that electron scattering in metals is quantum mechanically controlled by the same paradigm that is responsible for other electronic properties of materials. Realizing this is central to developing a modern understanding of the origins of the Wiedemann–Franz law. The Sommerfeld relationship points the way to this understanding. Sommerfeld's equation was derived using only electronic wave functions. The relationship between electrical and thermal conductivity in Equation (6.1) implies that electronic wave functions harbor the source of electron scattering. Broad general knowledge of partial wave scattering [6] shows that the details of an approach to electron scattering in metals have already been developed. Establishment of a literature on molecular orbital calculation of electrical resistance in circuit components is underway [8].

In the logic of the Sommerfeld relationship, Fermi contact of electrons in s wave functions, is the largest source of thermalization of electrical energy in metals. Fermi contact is well known in metals from the contact shift portion of the Knight shift. Fermi contact was discovered in 1930 by Enrico Fermi in his studies of atomic spectroscopy [9]. This access point to understanding resistivity in metals appeared after Bloch's approach to the problem of electron scattering had been launched, and at the time it was not recognized as a direct route to understanding resistivity. As a consequence, the impact of electronic structure on resistivity and thermal conductivity in metals, required by the Sommerfeld relationship, has yet to be fully appreciated. Fermi contact may be best known as the source of nuclear magnetic resonance nuclear spin-spin coupling constants. In these cases the spin of one nucleus couples with a chemically distinct nucleus via Fermi contact involving valence electrons. Extensive use of nuclear magnetic resonance spectroscopy in organic and biochemistry accounts for the substantial volume of the literature in this area.

Quantum Magnetoresistance, h/e^2

Quantum magnetoresistance has been known since 1980 through studies of the quantum Hall effect [10]. Quantum Hall experiments are typically conducted in a regime of zero electron scattering. Hall experiments determine the change in voltage at right angles to an electric current that flows in a conducting strip that is perpendicular to an external magnetic field. Change in voltage and resistance at right angles to the applied current in a Hall experiment is an exclusive effect of the external magnetic field. In the absence of the external magnetic field there is no Hall voltage and no Hall resistance.[*]

By reducing the scale of the conducting strip to a few atomic layers, it is possible to observe quantum jumps in potential when the experiment is done under conditions of very limited electron scattering.[†] Like the Sommerfeld relationship, the analysis that led to understanding the quantum of magnetoresistance was based on pure electronic wave functions with no involvement of the lattice [10]. In the quantum Hall regime [10] there is no electron scattering to produce the observed resistance quantum. Magnetoresistance quanta are relatively large [11].

$$\frac{h}{e^2} = 25,812.807\Omega \tag{6.3}$$

Magnetoresistance quanta arise in the quantum Hall regime as quantization of magnetoresistance, that is, resistance to electron motion arising from the external magnetic field. This resistance is also connected to the electron "effective mass."

ELECTROMAGNETIC QUANTA

Evidence for magnetoconductance quanta,

$$\frac{e^2}{h},$$

were first reported, to our knowledge, in experiments published by Komatsubara et al. [12]. Magnetoconductance is the reciprocal of magnetoresistance and differs by a factor of 2 from its electrical analog, the electric conductance quantum,

$$\frac{2e^2}{h}$$

(see Figure 10.1 and Figure 10.2). In classical physics conductance is the reciprocal of resistivity.

Electric conductance quanta, conventional value 7.7480917×10^{-5} siemens, S $(1/\Omega)$, were identified [13,14] a few years after the discovery of magnetoresistance quanta. Conductance quanta are relatively as small as resistance quanta are large. The electric conductance quantum is two times the value of the reciprocal of the quantum of magnetoresistance, called the von Klitzing constant (Equation [6.3]). At present there is substantial literature in the area of electric conductance quanta. The number of papers presenting observations of electric resistance quanta [14,15] is small (see

---

[*] For an introduction to the Hall effect see Chapter 9.
[†] In practice, this means very low temperatures, usually at or below the superconducting T_c for the material.

TABLE 6.1
Electromagnetic Quanta in Two Dimensions

Term	Name	Value	Reference
$\dfrac{h}{e^2}$	Magnetoresistance[1]	25,812.807 Ω	10
$\dfrac{e^2}{h}$	Magnetoconductance	38.740458 μS[2]	12
$\dfrac{h}{2e^2}$	Electric resistance	12,906.404 Ω	14, 15
$\dfrac{2e^2}{h}$	Electric conductance	77.480917 μS[3]	14

[1] Von Klitzing constant.
[2] Not yet officially recognized value.
[3] Micro-siemens, not yet officially recognized value.

Table 6.1). This dearth of reports may just be related to the small number of reports in general.

There is some confusion in the literature because of the existence of both electrical and magnetic quanta of resistance and conductance. The two systems are experimentally and theoretically distinct. They are easily confused for examples of the same phenomenon. We have grouped the quanta by term and name as two distinct groups. The name designations are simply placeholders. The term entries in the tables in this chapter come from the experimental literature.

Theoretical foundations for magnetoresistance and magnetoconductance quanta are found in the Landau quantization of an electron gas in a magnetic field [16]. Electric resistance and electric conductance quanta have a factor of 2 that is not present in the magneto-analogs. Theoretical understanding of this class of electromagnetic quanta is founded in the Landauer–Büttiker edge transport theory [17,18].

Magnetoresistance literature is primarily centered in the area of the quantum Hall effect. Experimental evidence of magnetoconductance quanta can be seen in the 1972 paper by Ando et al. [12] (see Chapter 10).

Electric resistance quanta have been studied in point contacts, ballistic wire, and nanotube experiments. A 2001 paper by de Picciotto et al. [15] gives data on both electric conductance and resistance quanta in a mix of two- and four-terminal experiments with a ballistic quantum wire at a temperature of 300 mK. Gallium arsenide/aluminum gallium arsenide, GaAs/AlGaAs, heterostructure nanowires were utilized under conditions of essentially zero electron scattering to study resistance and conductance as a function of gate voltage in these experiments. The conductor wires were prepared with closely controlled geometry by epitaxial growth onto the cleaved surface of a high-quality GaAs/AlGaAs heterostructure. The resistance quantum corresponded roughly to $h/2e^2$ and apparently arose from the contacts. The wire was shown to be essentially free from manifestations of resistance.

TABLE 6.2
Electromagnetic Quanta in Field-induced Phase Transitions

Term	Name	Value	Reference
$\dfrac{h}{0.5 \cdot e^2}$	Magnetoresistance	51,625.62 Ω [1]	19
$\dfrac{0.5 \cdot e^2}{h}$	Magnetoconductance	19.370229 μS [2]	20
$\dfrac{h}{4e^2}$	Electric resistance	6,453.202 Ω [1]	21, 22
$\dfrac{4e^2}{h}$	Electric conductance	154.961834 μS [2]	21, 22

[1] Not yet officially recognized value.
[2] Micro-siemens, not yet officially recognized value.

Literature for electromagnetic quanta in field-induced phase transitions, Table 6.2, is sparse as of yet. Reports thus far have been limited to three systems. These systems are (1) magnetic field-induced insulator–quantum Hall–insulator transitions in a semiconductor Hall bar system [19]; bilayer graphene [21]; and epitaxially prepared cuprate two dimensional devices [22] (see Chapter 10). Electromagnetic quanta in field-induced phase transitions can be associated with chiral electrons in parallel electric and magnetic fields, as pointed out by Novoselov et al. [21]. Chiral electrons, as they describe them, are intrinsically four-dimensional.[*] In the bilayer graphene, the electric field on the z-axis is created by biexcitonic Bose–Einstein correlates [23,24]. These electrically neutral charge transfer structures, in which symmetric radical anion–radical cation pairs are formed in the two layers of bilayer graphene, polarize the surfaces. This polarization results in the presence of an electric field in the z-axis. In the experiments described by Bollinger et al. [22], the chirality-producing z electric field was intentionally applied in the quantum Hall regime (see Chapter 10). Chiral electrons were not involved in the phase transitions observed by Hughes et al. [19]. The magnetoresistance/magnetoconductance quanta that they investigated are included here because of their involvement in both directions of a Mott transition in the quantum Hall regime. The experiments of Hughes et al. [19] carefully explored the transition to and from the quantum Hall state from and to the insulator state as a function of magnetic flux density and temperature.

Table 6.3 introduces electric and magnetic quanta in two dimensions. This is where the modern quantization of atomic level mechanics started. The Josephson constant, was the first of the quanta in the tables in this chapter to appear in the

[*] The experiments of Novoselov et al. [21].did not involve a phase transition but involved the same quanta as seen in Ref. 22, which followed five years later. It is for this reason that the 2006 experiment is discussed here.

TABLE 6.3
Voltage and Magnetic Flux Quanta

Term	Name	Value	Reference
$\dfrac{h}{e}$	Magneto magnetic flux	4.13566728 pWb[2]	25
$\dfrac{h}{2e}$	Electric magnetic flux	2.06783364 pWb[2]	26
$\dfrac{e}{h}$	Magneto voltage	241,798.935 Hz•V^{-1} [3]	27
$\dfrac{2e}{h}$	Electric voltage[1]	483,597.87 Hz•V^{-1}	28, 29

[1] Josephson constant.
[2] Pico–Weber, not yet officially recognized value.
[3] Not yet officially recognized value.

literature. Voltage standards are based on the Josephson constant [28], Equation (6.4).* This constant can be observed in

$$K_j = \frac{2e}{h} \tag{6.4}$$

which is the Josephson effect when two superconductors are connected by an insulating bridge that permits quantum tunneling. Superconducting quantum interference devices (SQUIDs), are an important technical advance resulting from the Josephson junction.

Table 6.3 highlights the differences between the electromagnetic quanta that are generated on the electric side and the corresponding quanta that are generated on the magnetic side of experimental physics. The two sets of quanta differ by a factor of 2 (or ½) for matched pairs; for example, what we refer to as *magneto voltage* and *electric voltage* quanta, respectively, in Table 6.3. The factor of 2 appears to come from the differences between the fundamentals of the Landau levels [16] that are the foundation for our understanding of the quantization of magnetoresistance, and the Landauer [17], Büttiker [18], edge transport theory foundations of our understanding of the quantization of the electric side of the problem.

The magneto voltage quantum [27] was discovered not long after the Josephson constant in extensive studies of the alternating current Josephson effect [28,29]. Electric magnetic flux quanta are the reciprocal of the Josephson constant, h/e^2, $2.06783364 \times 10^{-15}$ webers, Wb. Both of these quanta arise on the electric side of

---

* The units in Equation (6.4) are V^{-1}s^{-1} (frequency divided by voltage). These quanta were discovered in the study of electron tunneling between two superconductors connected by a nonsuperconducting bridge, for example, tin oxide, SnO [28].

quantum generation and have the factor of 2 that is missing from magneto magnetic flux and the magneto voltage quanta.

The final section in the discussion of the origins of resistance is known more generally as electron effective mass. In condensed matter physics, broad, routine use is made of the tools for molecular quantum mechanics, in which the mass of the electron has its standard value. Data suggesting an intimate connection between electron "effective mass" and magnetoresistance will be pointed out as it is encountered, later in this book.

DISSIPATIVE MAGNETORESISTANCE

When a metal is placed in a magnetic field it is possible to observe cyclotron resonance of conducting electrons in the metal. Cyclotron motion of electrons in conduction bands is caused by interaction between the thermal momentum of the negatively charged electrons in the conduction band and the applied magnetic field. The classical description for cyclotron motion of an electron is given by

$$m_e \omega_c = eB_o \qquad (6.5)$$

The cyclotron frequency, ω_c, is the ratio of the classical velocity of the electron to the radius of the cyclotron orbit. When the experiment is conducted, utilizing the thermal momentum of the electron, obtained by use of the Boltzmann equation, the product of m_e and ω_c does not equal eB_o. The equation is justified by introducing m_e^*, electron "effective mass" to replace m_e. No one thinks that the mass of the electron changes. That is why the word "effective" is used.

Thermal electrons are not supposed to experience any resistance to their motion. There is no resistivity for a thermal electron in a quantum mechanical system where the electron is not exposed to a differential in accelerating field(s). Thermal electrons in a magnetically induced cyclotron are subjected to a centripetal acceleration by the magnetic field. So, cyclotron electrons are subject to resistivity. Even though there is no electric field to drive their motion, there is a magnetic field to drive their motion. The "effective" loss of mass for the electron is just a bookkeeping device for the electron momentum that generates the cyclotron frequency. The electron momentum is assumed to be generated by a classical thermal velocity equilibrium from $k_B T$ and the electron effective mass. The phenomenological problem is that the momentum of the electrons in cyclotron resonance is smaller than the momentum expected from the classical thermal velocity. The observation is that electrons in a magnetic field behave as if they had an unexpectedly low momentum. This observation is valid. Using the description, electron "effective mass" sidesteps the question of the cause of the effect. The problem is the electron "classical velocity." Our analysis suggests that the observed electron momentum decrease is caused by dissipative electron scattering due to the presence of the magnetic field and the cyclotron oscillator that drives the cyclotron current. This scattering of thermal electrons dissipates the information associated with the conducting electron moving in a cyclotron orbit. That increase in entropy will decrease the effective temperature (velocity) of the electron (see Chapters 8, 9, and 10). Dissipative thermal effects caused by magnetism are not new to this field [30]. Refrigerative resistance effects represent a new understanding of the magnetoresistance process operating without additional fields. In this model the observation of the effect of magnetoresistance requires both an external magnetic field and the cyclotron oscillator. Without the oscillator there will be no current. Any current that might begin will be quenched by the magnetoresistance thwarting the possibility of observation of effects.

Developing an understanding of dissipative magnetoresistance is the fundamental requirement for understanding electron effective mass in metals. For a classical treatment of dissipative magnetoresistance, see Equation (9.3). Magnetoresistance is best known in a form coupled with electrical resistance as "Hall resistance," in Hall effect experiments. See the discussion of a classical Hall experiment in Chapter 9. Hall resistance is orthogonal to the voltage that drives the electric current in the Hall experiment, though the voltage-driven momentum of the electrons creates the observed magnetoresistance. Kapitza made extensive studies of magnetoresistance in the first part of the twentieth century [31]. Kapitza's efforts at mathematically describing the observed increase in electrical resistance with applied magnetic field were significantly challenged by the fact that the electric and magnetic variables in his experiments were conflated rather than partially isolated as is possible in the Hall experiment. Magnetoresistance is ultimately the source of the apparent lost electron momentum in the problem of electron effective mass. It is also the source of the Hall effect and the quantum Hall effect.

Effects of magnetoresistance quanta are observed in the quantum Hall regime where there is no scattering of electrons and no electrical resistance due to electron scattering. Under these conditions, electron scattering cannot be the source of the magnetoresistance. Quantum magnetoresistance is integral to the architecture of the quantum mechanics of electricity and magnetism. It appears that resistance is intimately associated with electron acceleration caused by a magnetic field and will manifest in quantum mechanical systems as partial wave scattering in systems composed of atoms. In the quantum Hall regime, magnetoresistance manifests as a quantum of magnetoresistance. Quanta of electric resistance manifest in systems that contain point contacts but do not manifest partial wave scattering of electrons [32]. Conductance quanta that correspond to the reciprocals of magnetoresistance and electric resistance quanta appear under the same generative experimental conditions.

QUANTUM MECHANICS AND THERMAL CONDUCTANCE

This chapter started with the Sommerfeld relationship between electrical conductivity and thermal conductivity. It is appropriate to end the chapter with an up-to-date view of the quantum mechanics of pure thermal conductivity, that is, thermal conductivity that is not mediated by electrons. The Sommerfeld relationship considers thermal conductivity that is mediated by electrons and provides the proportionality constant, the Lorenz number, with electrical conductivity. Recent publications on quantum thermal conductance include a report of experimental measurement of the quantum of thermal conductance [33] and two earlier reports concerning the transport theoretical foundation for the observation [34,35]. For ideal coupling between hot and cold reservoirs by a ballistic thermal conductor, theory gives the quantum of thermal conductance [33–35], Equation (6.6), as a component of the thermal conductance.

$$g_0 = \frac{\pi^2 k_n^2 T}{3h} \tag{6.6}$$

A numerical value for the quantum of thermal conductance is, $g_0 = (9.456 \times 10^{-11}$ W/K$^2)T$. The critical measurements in the experiment reported by Worlock et al. [33] were made at temperatures below 100 mK. For the temperatures that are central to the Wiedemann–Franz law and the Sommerfeld relationship, thermal conductance quanta would not be detectable in the thermal noise. These quanta, along with phonon effects, will certainly be there. Detectability for these small effects will always be a problem.

Before we move on to a more detailed discussion of the Hall effect, related topics, and the details in the landscape surrounding superconductivity, it seems reasonable to quickly review some of the fundamentals of quantum mechanics and condensed matter chemistry and physics that we will need in our discussions. Metals and insulators, and electron transport are the root subjects of the following two chapters.

REFERENCES

1. A. Sommerfeld, *Die Naturwissenschaften*, 1927, *15*, 825–832.
2. R. Franz, G. Wiedemann, *Ann. Physik*, 1853, *165(8)*, 497–531.
3. L.Lorenz, *Ann. Phys.*, 1872, *147*, 429.
4. G. S. Kumar, G. Prasad, and R. O. Pohl, *J Materials Sci.*, 1993, *28*, 4261–4272.
5. J. J. Sakurai, *Modern Quantum Mechanics*, revised ed., S.F. Tuan, Ed., 1994, Addison-Wesley, Longman, New York, 7.6 Method of Partial Waves, pp. 399–409.
6. See, e.g., M. Marder, *Condensed Matter Physics*, Second Edition, 2010, J. Wiley, New York: (a) Drude Model, p. 453; (b) Bloch wave vector, p. 182.
7. F. Bloch, *Proc. R. Soc. London A*, 1980, *371*, 24–7. The quotation is the fourth full paragraph in this paper, starting at the bottom of p. 24.
8. See, e.g., Y.-M. Lin, V. Perebeinos, Z. Chen, and P. Avouris, *Phys. Rev. B*, 2008, *78*, 161409(R) 1–4.
9. E. Fermi, *Nature*, 1930, *125*, 16–17.
10. K. von Klitzing, G. Dorda, and M. Pepper, *Phys. Rev. Letters*, 1980, *45*, 494–497.
11. Conventional value of the von Klitzing constant, National Institute of Standards and Technology, http://physics.nist.gov/cgi-bin/cuu/Value?rk90.
12. T. Ando, Y. Matsumoto, Y. Uemura, M. Kobayashi, and K. F. Komatsubara, *J. Phys. Soc. Japan*, 1972, *32*, 859.
13. B. J. van Wees, H. van Houten, C. W. J. Beenakker, J. G. Williamson, L. P. Kouwenhoven, D. van der Marel, and C. T. Foxon, *Phys. Rev. Letters*, 1988, *60*, 848–850.
14. D. A. Wharam, T. J. Thornton, R. Newbury, M. Pepper, H. Ahmed, J. E. F. Frost, D. G. Hasko, D. C. Peacockt, D. A. Ritchie, and G. A. C. Jones, *J. Phys. C*, 1988, *21*, L209–214.
15. R. de Picciotto, H. L. Stormer, L. N. Peiffer, K. W. Baldwin, and K. W. West, *Nature*, 2001, *411*, 51–54.
16. L. D. Landau, *Z. Phys. A*, 1930, *64*, 629-37; for an English translation, see, D. ter Haar, Ed., *Collected Papers of L. D. Landau*, 1965, Gordon and Beach, London; pp. 31–38.

17. R. Landauer, *Philos. Mag.*, 1970, *21*, 863.
18. M. Büttiker, *Phys. Rev. Letters*, 1986, *57*, 1761–1764.
19. R. J. F. Hughes, J. T. Nicholls, J. E. F Frost, E. H. Linfield, M. Pepper, C. J. B. Ford, D. A. Ritchie, G. A. C. Jones, E. Kogan, and M. Kaveh, *J. Phys.: Condens. Matter*, 1994, *6*, 4763–4770.
20. *Ibid.,* Figure 2, p. 4765.
21. K. S. Novoselov, E. Mccan, S. V. Morozov, V. I. Fal'ko, M. I. Katsnelson, U. Zeitler, D. Jiang, F. Schedin, and A. K. Geim, *Nature Phys.*, 2006, *2*, 177–180.
22. A. T. Bollinger, G. Dubuis,, J. Yoon, D. Pavuna, J. Misewich, and I. Božović, *Nature*, 2011, *472*, 458–460.
23. J. P. Eisenstein, and A. H. MacDonald, *Nature*, 2004, *432*, 691–694.
24. S. A. Moskalenko, and D. W. Snoke, *Bose-Einstein Condensation of Excitons and Biexcitons*, 2000, Cambridge U. Press, New York.
25. F. E. Camino, W. Zhou, and V. J. Goldman, *Phys. Rev. B*, 2005, 155313, 1–6.
26. C. E. Gough, M. S. Colclough, E. M. Forgan, R. G. Jordan, M. Keene, C. M. Muirhead, A. I. M. Rae, N. Thomas, J. S. Abellt, and S. Surtont, *Nature*, 1987, *326*, 855.
27. B. N. Taylor, W. H. Parker, and D. N. Langenberg, *Rev. Mod. Phys.*, 1961, *41*, 375–496.
28. B. D. Josephson, *Phys. Letters*, 1962, *1*, 251–253.
29. B. D. Josephson, *Rev. Mod. Phys.*, 1974, *46*, 251–254.
30. See, e.g., J. G. Daunt, and C. V. Heer, *Phys. Rev.*, 1949, *76*, 715–717. This paper reported the T_c for titanium, 0.53 K, observed by use of magnetic refrigeration.
31. P. L. Kapitza, *Proc. Roy. Soc. A*, 1929, *123*, 292–341.
32. See, e.g., K. Sekiguchi, A. Yamagachi, H. Miyajima, A. Hirohata, and S. Usui, *Phys. Rev. B*, 2008, *78*, 224418(1–5). This paper deals with nickel magnetic point contacts.
33. K. Schwab, E. A. Henriksen, J. M. Worlock, and M. L. Roukes, *Nature*, 2000, *404*, 974–977.
34. L. G. C. Rego, and G. Kirczenow, *Phys. Rev. Letters*, 1998, *81*, 232–235.
35. D. E. Angelescu, M. C. Cross, and M. L. Roukes, *Superlatt. Microstruct.*, 1998, *23*, 673–689.

7 Metals and Insulators

One of the central questions in all materials is: What is it that holds the material together and is responsible for its properties? The simple question of what holds hydrogen together has been a standard at undergraduate oral exams in chemistry and physics for 50 years. The bond in hydrogen gas, H_2, is remarkable in that spin pairing[*] for the two electrons generates a bond that lowers the enthalpy of two hydrogen atoms by 436 kJ for every mol of H_2 formed. Hydrogen gas is an insulator, and the bond that holds the molecule together is a covalent, electron-pair bond.

Metals are characterized as having massively degenerate, singly occupied, three-dimensional wave functions that formally extend to the edges of the phase. As the example of lithium bonding from small clusters to bulk metal shows,[†] metallic bonding occurs when the free energy decrease associated with electron delocalization makes delocalized bonding energetically favorable compared to electron pair bonding. This occurs when covalent bonding and delocalized metallic bonding differ by the order of k_BT. The phase transition that occurs in the bonding shift is known as a Mott transition [1]. Mott transitions are also known as metal–insulator transitions and are very familiar in condensed matter physics and chemistry [2]. Superconductor–insulator transitions[‡] are also documented in the literature [3,4]. On the insulator side of the transition, the bonding in the material is covalent, electron pair bonds, and the material is not a conductor. Mott transitions are known to occur as a function of temperature [1,2], pressure [5], magnetic field [3], and electric field [4].

Metallic bonding forms some of the strongest chemical bonds known. Tungsten, W, a transition metal, has a melting point and boiling point at one bar of 3680 and 5828 K, respectively [6]. Rhenium and tungsten are the most refractory of the elements, which dramatically demonstrates the bond strength at extremely high temperatures in these metals. Tungsten, rhenium, and other refractory metals are held together by exceptionally strong chemical bonds that are delocalized, one-electron bonds.

Treatment of bonding in metals from a molecular orbital perspective in the physics community has focused on band theoretical approaches to the subject. Band theory is a qualitatively different approach to chemical bonding in materials than approaches based upon the enthalpy of bond formation. Different factors are important in determining the binding enthalpy of atoms when you compare bonds formed with one electron per orbital, metals, or two electron per orbital insulators. Because of the differences, the binding enthalpy of metals receives much less attention than

---

[*] Spin-paired electrons refers, in this context, to two electrons in the same wave function with opposite spins. Constructive interference between the two atomic basis functions is a major feature of the bonding in hydrogen gas. The magnetic pairing interaction is much smaller and has the same sign.

[†] See, Bonding in Small Clusters of Alkali Metals in this chapter.

[‡] The transitions cited were observed in very thin films in the quantum Hall regime. Bollinger et al. referred to the transition they reported [4] as a superconductor–insulator transition.

it does in materials that are bonded by electron-pair bonds. Likewise the physical properties of the materials are radically different, for example, specific conductance, specific resistance, and ductility are exclusively in the metal domain. Chemists focused on the domain of small molecules give this area little attention.

Metallic bonding includes the effects of the astronomical number of magnetic interactions that occur in metals. The number of negative charge carriers in a typical metal is of the order of 10^{23} electrons/cm^3. Magnetic interactions between conducting electrons and, e.g., partially filled d- or f-basis electrons associated with adjacent atomic cores are certain to occur. See the discussion of refractory elements, Table 7.2.

One prominent feature that distinguishes metals is the so-called energy gap.[*] Insulators typically have such a gap between doubly occupied, nonconducting wave functions and vacant wave functions that are available for conduction using single electrons. Metals do not form an energy gap between occupied and vacant wave functions, but form conduction bands from the formally degenerate singly occupied and vacant wave functions on the atoms.

In comparison to metals, bonding in insulators is relatively simple. There is zero electron spin in the average insulator, like a diamond. Insulators, including H_2, are built up using electron pair bonds that are associated with antibonding wave functions that create an energy gap between filled and vacant orbitals. Gilbert N. Lewis was one of the first prominent scientists to recognize the importance of the electron pair bond and its capacity to hold materials together and keep individual materials distinct [6]. In 1916, Lewis recognized that formation of electron pairs produced either molecules, stable atoms (column 18 of the periodic table, the rare gases), or stable ions. It is, of course, possible to form electron-paired states that are unstable. These states are only of concern in studies of transient phenomena. A very readable historical account of the electron-pair bond has been produced by Shaik and Hiberty [7].

Worldwide chemical activity adds roughly 12,000 compounds to the Chemical Abstracts database each week. Most of those compounds are organic or biological compounds. A vast majority of those groups are insulators. Typically, these compounds are even electron, spin-paired molecules composed of atoms predominantly from the first three rows of the periodic table. The Chemical Abstracts database contains on the order of 5×10^7 compounds, the dominant majority of which fall into the organic or biological classification.

Chemists have focused so strongly on the importance of the two-electron chemical bond that the literature in chemistry on bonding in metals is by comparison quite thin. Valence bond theory [8], which is a valid quantum mechanical alternative to molecular orbital theory, generally makes no mention of bonding in metals such as tin, Sn.

Mesoscopic, singly occupied molecular orbitals are the feature that distinguishes metals from electron pair-bonded insulators and salts. These very large, diffuse, one-electron orbitals arise from linear combinations of singly occupied metal atomic orbitals in the metal lattice. It is a remarkable feature of quantum mechanics

---

[*] The insulator energy gap should not be confused with the superconductor energy gap, which is formed by a completely different mechanism, namely the loss of dissipative resistivity. See the discussion of the electronic heat capacity at the superconductor critical temperature, T_c.

that a pair of bonded Rh atoms in the lattice has a bonding network that involves approximately two electrons per rhenium. Yet the number of electrons that actually participate in the bonding is astronomical. Metal conduction bands are collections of degenerate or nearly degenerate orbitals, each containing a single electron, that formally extend over the entire lattice. In principle, the same process could happen when carbon atoms come together to form a single crystal of diamond, but it doesn't. Diamond is bonded by electron pair bonds and is an insulator with resistivity in the range 10^{16} to 10^{18} $\Omega\cdot$m, depending upon the source of the measurement.

FACTORS IN FORMING METALS OR INSULATORS

From a chemical perspective, there are at least two historically important conceptual physical factors that are involved in the subtle difference that makes tin a metal, and phosphorous a nonmetal. The first is the effective size of the valence level wave functions that are used by the element to form bonds. Generally, the more diffuse those functions are on an individual atom, the weaker the bonds that can be formed with other atoms. The effective size for valence level wave functions increases monotonically going down the periodic table, with increasing values of n for the valence level bond forming wave functions. The size of atoms decreases going from left to right along a row in the periodic table. Half filled subshells cause a decrease in atomic size and produce discontinuities in plots of atomic size against atomic number.

Nuclear charge seen by the valence-level bonding electrons is the second major factor. Nuclear charge experienced by valence electrons in atoms and molecules is not the same as the charge on the nucleus. Nuclear charge for lead, Pb, is 82; however, a $6p$ lead electron will not experience a charge of +82, because inner shell electrons screen the nuclear charge. Physicists and chemists are familiar with the Slater rules for screening constants. One empirical approach to evaluation of effective nuclear charge for an atom is electronegativity. Pauling and Mulliken developed the two widely used definitions of electronegativity. Pauling and Mulliken electronegativity are defined in the following paragraphs.

The difference between two Pauling electronegativity values, χ_A and χ_B, is defined by the dimensionless difference between the homonuclear and heteronuclear bond enthalpies (in eV) in valence bond theory.

$$\chi_A - \chi_B = (eV)^{-\frac{1}{2}}\sqrt{E_d(AB) - \left[E_d(AA) + E_d(BB)\right]/2} \tag{7.1}$$

Mulliken's definition of electronegativity is somewhat more physical, in that it is the scaled sum of the atomic ionization energy and electron affinity (Equation [7.2]) for energies in eV. The scaling constants place the Mulliken electronegativities on the same scale as those obtained by Pauling's bond enthalpy procedure.

$$\chi_A = 0.187(eV)^{-1}(IE_A + EA_A) + 0.17 \tag{7.2}$$

In Equation (7.2), IE_A is ionization energy, and EA_A is electron affinity; both are in eV.

TABLE 7.1

Periodic Table of the Elements Showing the Demarcation between Metals and Nonmetals As a Zigzag Dark Line from Boron, B, to Element No. 118, and Showing Pauling Electronegativity Values for the Elements

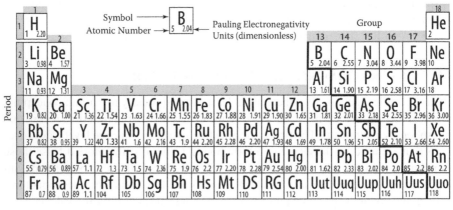

Source: L. R. Murphy, T. L. Meek, A. L. Allred, and L. C. Allen, *J. Phys, Chem. A*, 2000, 104, 5867–5871.

Table 7.1 is a periodic table of the elements showing Pauling electronegativity values [9,10]. You can see from Table 7.1 that two significant figures will capture any predictive value as far as metals and nonmetals are concerned. The dark zigzag line at the right of the table is the traditional demarkation between metals and nonmetals. Elements to the left of the zigzag line are all, at least marginally, metallic. Elements to the right of the same line are all at least marginally nonmetallic. Demarcation lines between metals and nonmetals are not crisp. The border near the zigzag line is a treasure trove of condensed matter special cases. Modern practice includes subcategories for both semiconductors and semimetals, both of which lie along the zigzag border [11]. A semimetal has a very small population of charge carriers per atom at temperatures near 0 K. Elemental semiconductors, by the physics definition, are found in column 14 of the periodic table; however, Sn and Pb, are generally classed as metals by both physicists and chemists.* Carbon has such a rich structural chemistry that organic chemistry is devoted to its study. At this time Chemical Abstracts lists more than 65 M sequences, and 43 M commercial chemicals. Carbon provides the structural basis for the vast majority of these substances.

---

* The low temperature allotrope of tin, α-tin, has a diamond crystal lattice and is a semiconductor.

Two factors in bonding that are important to the distinction between metals and nonmetals are atomic size and electron attracting power (electronegativity). For the nonradioactive elements, using these factors, cesium, Cs, is the extreme metal in the periodic table.* Fluorine, F, is the extreme nonmetal in the periodic table. Elements in the *d*- and *f*-transition series, columns 3–12, and elements 58–71 and 90–103 are all metals. This is because the highest energy electrons in transition element atoms are, at least formally, inner-shell electrons. Inner-shell electrons have little impact on either valence (bonding) orbital size, or electron-attracting power. The reason for this is that as inner-shell electrons, the electrons that are added in the *d* series, for example, have a smaller principle quantum number and on average are closer to the nucleus than the outer-shell electrons for the element.

Arrangement of the elements in the periodic table is a direct consequence of atomic quantum mechanics. It is not surprising, then, that the properties of the elements, including modes of bonding, are arranged by the periodic table. Our central question is what is the defining difference between the metals and the nonmetals in the periodic table?† To develop an answer to this question we will look at the available experimental data for bonding in metallic elements. We know that electron-pair bonds are strong bonds. Metallic bonds involve delocalization of electrons. Delocalization of electrons increases the intrinsic entropy associated with the particle and lowers the free energy of the system.

BONDING IN SMALL CLUSTERS OF ALKALI METALS

Group 1 in the periodic table contains the alkali metals and hydrogen. Alkali metals are at the left edge of Table 7.1. They represent extremes in the bonding properties that are characteristic of metals. We will look at this group for information of the differences between bonding in metals and insulators. Lithium, Li, is the first metal in the periodic table. At low pressures in a gas phase jet, lithium dimerizes to form the Li_2 ground state [12].‡ Dilithium, Li_2, is a singlet, though the dissociation energy of the molecule is roughly half that of H_2, based on the dissociation energy of Li_2^+, 124 kJ/mol [12]. The bond dissociation energy of H_2^+, obtained by use of fluorescence excitation spectroscopy was reported as 256 kJ/mol (21,379.36 cm^{-1}) [13]. The corresponding value for the hydrogen molecule obtained by the same method was 432 kJ/mol (36,118.11 cm^{-1}) [13]. The reason that the dissociation energy of H_2^+ is 40 kJ/mol larger than half the bond dissociation energy of H_2 is primarily due to zero electron repulsion in H_2^+.

Although hydrogen sits right above lithium in the periodic table, hydrogen is an archetypical nonmetal and bulk lithium is a metal. The reason for this difference is the relative bond strengths for the dimers and clusters of the two respective atoms. Strong electron-pair bonds produce insulators. Weak electron-pair bonds promote

---

* Cesium is the highest atomic number, nonradioactive, element at the far left of the periodic table.

† Anyone who has worked with semiconductors in columns 14 and 15 in Table 7.1 knows about the looseness of the distinction made by drawing the zigzag line between metals and nonmetals in the table. Placement of the line is somewhat arbitrary. Zigzag strips including semiconductors and semimetals are also used.

‡ The radially symmetric, electron-pair bond that holds dilithium together is referred to as a σ-bond.

formation of clusters even in univalent atoms. Weak electron-pair bonds also compete with the delocalized one-electron bonds of metals at the Mott transition boundary.* As we proceed down the periods in group 1 of the periodic table the properties of the atomic dimers become more metal-like with each step.

Lithium as a solid is exceptionally reactive. One interesting question is: "What is the Li cluster size where lithium switches from electron-pair bonding with an energy gap for the cluster to delocalized, single-electron, metallic-bond formation in the material?" The electron spin resonance spectra for argon matrix–isolated lithium clusters $^6Li_7$ and $^7Li_7$ have been reported and analyzed [14]. The $^6Li_7$ and $^7Li_7$ clusters were observed in different experiments, starting with lithium with the appropriate isotopic composition. The spectra showed a single unpaired electron, appropriate for the odd-electron cluster, Li_7. The implication is that Li_6, if it were stable, would be closed-shell. Studies of optical properties of larger lithium clusters demonstrated that a lithium cluster as small as Li_{137} is directly approaching the optical response seen in the plasmon mode of lithium metal surfaces [15]. Moving down the periodic table, the electron-pair bond in the dimeric metals decreases in strength with each step.

Metals are materials that will not support an internal electric field caused by an external applied electric field [16]. A corollary of this general statement is that a metal can not have an internal electric dipole moment. de Heer et al. used this principle in their studies of the dipole moments of small sodium, Na, clusters [17]. The largest dipole moment observed was that for the sodium trimer, Na_3 [17]. In their conclusion the authors state:

> ... the electric deflection measurement discussed here gives a comprehensive picture of the response of small sodium clusters to static electric fields. The nearly vanishing electric dipole moments, even for clusters as small as the sodium trimer, demonstrates that the electric fields surrounding alkali clusters are very small, as expected for a classical metallic object [18].

The measured dipole moment for Na_3 clusters was reported as ~0.01 D, debye [17].

The highest atomic number for a nonradioactive alkali metal is held by cesium, $_{55}Cs$. Very cold (~300 μK) dicesium, Cs_2, was formed by photoassociation of Cs vapor in high vacuum and shown to have an electronic triplet state [19]. This means that no energy gap is formed on making a bond between two cesium atoms. There is no energy gap for Cs_2, which means that Cs_2 is among the smallest examples of a metal, though it lacks the macroscopic properties of a metal.

BONDING MODELS FOR INSULATORS

The known bond enthalpy of Li_2^+ suggests an enthalpy over 200 kJ/mol for Li_2. This would mean that dilithium is a respectable electron-pair bonded diatomic molecule. For comparison, the respective bond enthalpies of fluorine, F_2, and chlorine, Cl_2, gas are 159 and 242 kJ/mol [20]. Both fluorine and chlorine are dielectric materials,

---

* For a discussion of Mott transitions, see the discussion under Mott Transitions in Metallic Systems in this chapter.

though they are both exceptionally chemically reactive. The change in bonding between Li_2 and Cs_2 follows the general trend of decreasing homonuclear bond strength with increasing principal quantum number. The bond enthalpy comparison between F_2 and Cl_2 is an exception. Bond enthalpy generally decreases going down the periodic table, but because of the extreme electronegativity of fluorine, F_2 has an unusually weak bond.

Although Li_2 has a band gap, bulk lithium is most assuredly a metal. It appears that metals are formed when the bond enthalpy of an atomic cluster is greater for bonding with delocalized electrons in singly occupied orbitals than it would be for formation of electron-pair bonds. This understanding of one-electron metallic bonds should help us to expand our appreciation of bonding in metals.

For insulators, the model we use has been presented countless times in the past. Bond formation to make H_2 is just the bond enthalpy that develops when two hydrogen atoms approach each other and form a bonding and antibonding pair of wave functions from a pair of atomic $1s$ wave functions. The bonding process is a first-order perturbation to the atomic system, (see Figure 7.1).

Perturbation theory for chemical bonding was an early development [21]. It was based on the electron-pair chemical bond that has an energy gap. The bond in molecular hydrogen is an archetype of this type of bonding. The perturbation expansion assumes that the binding energy of the two hydrogen atoms with each other is small compared to the energy required to dissociate the molecule into protons and electrons.

It does not seem likely that a more detailed model than first-order perturbations will be of much use for building an understanding of bonding in the hydrogen molecule or other insulators. At present, bonding is largely dealt with using sophisticated

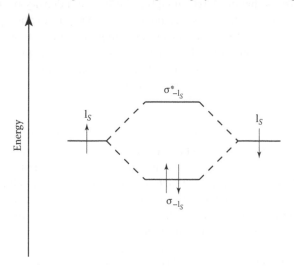

FIGURE 7.1 Bond formation between two hydrogen atoms. A first-order perturbation that forms an energy gap. (*Note:* The σ refers to the radial symmetry around the bond axis of the bond between the two protons. The $\sigma_{1s}{}^*$ refers to the antibonding wave function that forms between the two atoms. The gap between the bonding and the antibonding orbital is the energy gap of insulators.)

self-consistent quantum mechanical models, and the working applications of perturbation theory involve highly specific problems. The important features of Figure 7.1 are the vacant anti-bonding wave function; the energy gap between the bonding wave function and the antibonding wave function; and the energy state associated with bond formation.

TOWARD A QUALITATIVE MODEL FOR BONDING IN METALS

The cesium vapor experiment discussed above demonstrated that the ground state of dicesium, Cs_2, is very weakly bonded, degenerate triplet, and metal-like [19]. There apparently is zero first-order bonding interaction when two cesium atoms come together. The dimer starts out in a metal-like state. It is not likely from this start that the formalism for insulators will work here. The existing formalism for bonding works well for systems bonded with electron-pair bonds that have an energy gap. This does not include cesium clusters or Cs metal.

In insulators it is typical that the even-electron states have a total electron spin of zero. The negative enthalpy associated with spin pairing can be thought of as a decrease in the interelectron repulsion. This effect contributes to the increased electron density between the bonded nuclei when the electrons are in the bonding wave function in Figure 7.1. Since the Coulomb potential is ultimately the force that holds bonded atoms together, an increase in electron density between the nuclei increases bond strength. It appears that the feature of metals that performs a similar function of increasing electron density between bonded nuclei is also a magnetic coupling but not spin pairing.

In metal clusters that are large enough to have a high density of unpaired spins, not only are their many electron spins, but the electron spins explicitly participate in chemical bonding. All of the questions of coupling of angular momentum for spins and orbitals must be considered. Looking at the position in the periodic table of the refractory metals* it is possible to see that early d-series metals are often refractory (see Table 7.2). As the principal quantum number increases, bond strength in a given group generally decreases so the lanthanides and actinides are only represented by thorium, Th.

Nonmetals in Table 7.2 only include carbon and boron. Although superconductors are based on both of these elements in association with other elements, the bonding in elemental carbon and boron is electron-pair bonding, and the materials are fundamentally insulators. To be sure, β-graphite, the elemental thermodynamic standard state for carbon, is a conductor in two of three axes. The conduction in this case is metallic conduction with individual electrons moving in nonbonding or antibonding π-symmetry wave functions of the graphite.

---

* Standard definitions of refractory metals often include elements selected from columns 4 through 9 of the periodic table. The general definition of a refractory metal is an engineering question dealing with wear resistance and resistance to thermal effects in commercially important metals. Radioactive metals or precious metals are not generally included in the engineering definition. The definition of refractory used here and in Table 7.2 is a one-bar vaporization temperature "boiling point," b.p., of 4200 K or above. Vaporization temperature provides a crude index to bond enthalpy at the surface of the liquid metal.

TABLE 7.2

Refractory Metals and Nonmetals

Z	Sym	Name	Group	Period	M.P. (K)	B.P. (K)
75	Re	Rhenium	7	6	3453	5869
74	W	Tungsten	6	6	3680	5828
73	Ta	Tantalum	5	6	3269	5731
76	Os	Osmium	8	6	3300	5285
43	Tc	Technetium	7	5	2473	5150
90	Th	Thorium	f	7	2028	5061
41	Nb	Niobium	5	5	2741	5017
42	Mo	Molybdenum	6	5	2890	4912
72	Hf	Hafnium	4	6	2500	4876
77	Ir	Iridium	9	6	2716	4701
40	Zr	Zirconium	4	5	2125	4682
44	Ru	Ruthenium	8	5	2523	4423
6	C	Carbon	14	2	3948	4300
5	B	Boron	13	2	2573	4200

Source: http://www.webelements.com/

Engineers who deal with wear are more concerned with stability of the solid than they are with chemical bond enthalpy. As a consequence, the engineering tables are arranged by decreasing melting point. A change to that basis would move thorium from sixth position to fourteenth, which illustrates how crude these approximations to bond enthalpy order are. Nonetheless, it seems fair to suggest that the strongest homonuclear chemical bonds are in all probability the bonds that hold metals like rhenium, Rh, and tungsten, W, together. These are one-electron bonds that are augmented by magnetic interactions between valence electrons (conduction band electrons) and other unpaired electrons associated with the atomic cores.

One of the things that holds metals together, for example, Cs_2, is delocalization of the valence-level electrons in the system over the atoms. There is a fundamental entropy advantage here that will be limited in nanoscopic clusters and the metal by a correlation length analogous to the Ginzberg–Landau electron correlation length.

Cs atoms, atomic mass 132.9 u, amu,* are much more strongly bound in the solid state than the nearest lower-mass element xenon, Xe, atomic mass 131.2 u, which has one less electron per atom. Respective one-atmosphere boiling points for these two elements are 944 and 165.3 K [20,22]. The almost 779 K difference in the two boiling points reflects the cohesion enthalpy of cesium metal as a liquid at high temperature compared to the cohesion enthalpy of xenon as a liquid at a temperature below the boiling point of liquid nitrogen.

An orbital schematic for bonding in Cs clusters is suggested in Figure 7.2. Energy levels are shown for only two electrons in Cs_2, as there are no antibonding higher

---

* Atomic mass unit, symbol, u, 1/12 the mass of a nonbonded ^{12}C atom, also referred to as dalton, Da, or amu.

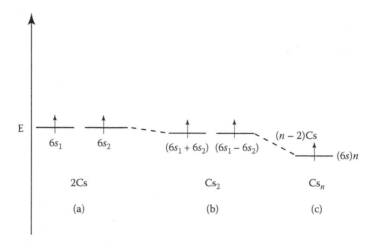

FIGURE 7.2 Highest occupied orbitals for bonding in Cs clusters: (a) 2 Cs, (b) Cs_2, (c) Cs_n. Only the highest energy electrons in the atoms, dicesium, and a cesium cluster are shown. In (c) the cluster MOs will be spread in energy to form a band. Note that energy gaps between (a) and (b) and between (b) and (c) are not to scale.

energy wave functions. There is no energy gap for Cs_n regardless of the magnitude of n. All of the subvalence level electrons in cesium clusters are in closed atomic inner shells and subshells, all of which are spin-paired, doubly occupied. The small decrease in energy of Cs atomic nonbonding wave functions in Cs_2 is due to delocalization of the two electrons over the two Cs cores rather than only one. This thermodynamic reduction of free-energy is generated by electron delocalization. Delocalization of electrons decreases electronic entropy as a function of temperature. Building a large cluster of Cs atoms will produce a system in which there is no gap between the half-occupied molecular wave functions formed from the half-filled $6s$ orbitals of Cs atoms and the vacant $6p$-basis orbitals of the same atoms. Starting with two independent atoms, Figure 7.2a, the binding energy is exceptionally small for Cs_2, Figure 7.2b. The binding energy per bond increases as the size of the cluster increases because of effects of electron delocalization. The energy for the conduction band in the Cs_n cluster in Figure 7.2c depends on the cluster size, below a minimum size for a sample of Cs metal.

Writing out a formal bonding expansion, including electron delocalization and all of the possible magnetic couplings, may be a necessary step for development of a deeper understanding of bonding in metals.

In electron spin-paired metallic systems, like Li_2, as the size of the even electron cluster increases, the energy gap between the highest occupied bonding orbital and the lowest energy vacant orbital steadily decreases until, at a cluster size that depends upon the metal, the system forms a conduction band made up of singly occupied molecular orbitals that are immediately adjacent to the lowest energy vacant orbitals of the system as suggested for Cs in Figure 7.2b. This is the point where the thermodynamics of electron delocalization is more favorable than first-order electron pair chemical bond formation. The cluster size for the transition to metallic behavior is a subject for both experimental and theoretical effort.

For lithium clusters, argon-matrix isolated Li_7 clusters gave electron spin resonance, ESR, spectra that were consistent with a single electron spin in the system [23]. Lithium clusters as small as Li_{139} gave electronic spectra that were interpreted as being consistent with significant plasmon mode electron delocalization in the cluster [24].* Metal cluster systems have the potential to display a size dependent Mott transition, where the transition from insulator to conductor is cluster size-dependent.

Wigner and Seitz were among the first to apply MO theory to the study of metal properties. Their 1933 efforts were heroic in view of the computational labor that was required [25]. Objections to the use of a standard LCAO MO approach were valid during the Landau era of condensed matter physics, because the necessary computing power just was not there. Advances in computational power since that time have significantly reduced these objections.

MOTT TRANSITIONS IN METALLIC SYSTEMS

Mott transitions are second-order phase transitions in metal/insulator systems that switch from metal to insulator at a given temperature and pressure. Switching involves changing from a metal with single-electron delocalized bonds to an insulator with electron-pair bonds and an energy gap. At the transition point for a given system, the two types of bonding must be in thermodynamic balance. Mott realized the possibility of these transitions when he was working on semiconductors, which are at the boundary between metals and insulators. Requirement of balanced metal and insulator systems at the transition are stringent. It is remarkable that there are as many known Mott transition systems as there are. There was a lapse of some time between the prediction [1] and the observation of a sharp Mott transition [2].

Modern materials science techniques have produced a number of interesting examples of specific Mott transitions, see, for example, Figure 7.3. Figure 7.3 presents scanning tunneling microscopy, (STM), images at two temperatures for a 1/3 monolayer of tin on a germanium (111) surface. Tin is a metal, and germanium, Ge, in this example, is a semiconductor with resistivity 4×10^{-3} Ω•m [26].

It is interesting to note that the insulator phase, shown in Figure 7.3 appears to be the less structured of the two phases. The fact that the low-temperature phase, on the right in Figure 7.3, is the insulating phase was demonstrated by the original authors by use of angle-resolved photoemission spectra of the samples. These spectra demonstrated that the low-temperature phase developed a surface band gap, consistent with a corresponding level of bond localization in the phase [26]. In spite of the homogeneous appearance in the low-temperature phase in the STM image, the phase transition was fully reversible. Based on literature precedents, the authors suggest that in the flat phase, the tin atoms occupy T_4 lattice sites, making the flat phase at least as organized as the high-temperature phase. A decrease in electron delocalization,

---

* Plasmon modes are collective quantum modes of surface electrons in metals. They are responsible for many of the optical properties of metals, such as metallic sheen, luster, and color.

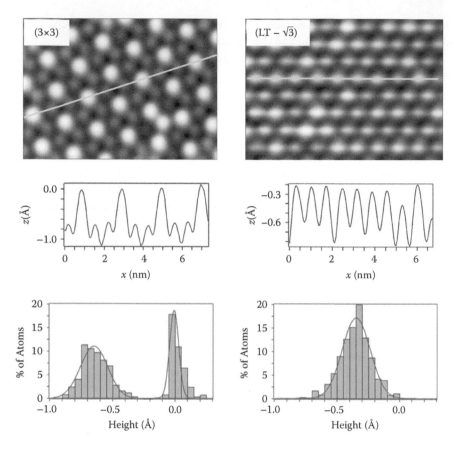

FIGURE 7.3 Top: STM image of (right) high-temperature phase, $T = 112$ K; (left) low-temperature phase, $T = 5$ K. Middle: Height profiles along the bright white lines of the images above. Bottom: Histogram distributions of the heights shown above. (From R. Cortés et al., *Phys. Rev. Letters*, 2006, 96, 126103 1–4.)

with development of an associated energy gap in the electronic band structure is the hallmark of Mott transitions.

Quantum mechanics controls the conducting electron spin in these systems. To understand the electron spin in a superconductor we must understand the quantum mechanical communication of electron spin throughout a metal. Communication of local electron spin response to external magnetic fields is an essential part of defining volume susceptibility and magnetic permeability for a magnetic domain in a metal. These topics are dealt with in the next chapter.

REFERENCES

1. N. F. Mott, *Proc. Phys. Soc.* (London) A, 1949, 62, 416–422.
2. See Mott's review, N. F. Mott, *Rev. Mod. Phys.*, 1965, 40, 877–883.

3. R. J. F. Hughes, J. T. Nicholls, J. E. F Frost, E. H. Linfield, M. Pepper, C. J. B. Ford, D. A. Ritchie, G. A. C. Jones, E. Kogan, and M. Kaveh, *J. Phys.: Condens. Matter*, 1994, 6, 4763–4770.
4. A. T. Bollinger, G. Dubuis,, J. Yoon, D. Pavuna, J. Misewich, and I. Božović, *Nature*, 2011, 472, 458–460.
5. P. Limelette, P. Wzietek, S. Florens, A. Georges, T. A. Costi, C. Pasquier, D. Jeŕome, C. Mézière, and P. Batail, *Phys. Rev. Letters*, 2003, 91, 016401 1–4.
6. G. N. Lewis, *J. Amer. Chem. Soc.*, 1916, 38, 762–785. http://www.webelements.com/
7. S. Shaik, P. C. Hiberty, *A Chemist's Guide to Valence Bond Theory*, 2008, Wiley-Interscience, Hoboken; chapter 1, pp. 1–14.
8. See, e.g., ibid. or G. A. Gallup, *Valence Bond Methods*, 2002, Cambridge University Press, Cambridge.
9. http://en.wikipedia.org/wiki/List_of_elements.
10. L. R. Murphy, T, L, Meek, A. L. Allred, and L. C. Allen, *J. Phys, Chem. A*, 2000, 104, 5867–5871.
11. M. Marder, *Condensed Matter Physics*, 2nd edition, 2010, J. Wiley, New York; see the periodic table inside the front cover, and pp. 279, and 283.
12. D. Eisel, and W. Demtröder, *Chem. Phys. Letters*, 1982, 88, 481–486.
13. A. Balakrishnan, V. Smith, and B. P. Stoicheff, *Phys. Rev. Letters*, 1992, 68, 2149–2152.
14. D. A. Garland, and D. M. Lindsay, *J. Chem. Phys.* 1984, 80, 4761–4767.
15. C. Bréchignac, Ph. Cahuzac, J. Leygnier, and A. Sarfati, *Phys. Rev. Letters*, 1993, 70, 2036–2039.
16. J. Jackson, *Classical Electrodynamics*, 3rd edition, 1998, Wiley, New York; pp. 27–29.
17. J. Bowlan, A. Liang, and W. A. de Heer, *Phys. Rev. Letters*, 2011, 043401 1–4.
18. Ibid., conclusion, p. 043401–4.
19. A. Fioretti, D. Comparat, A. Crubellier, O. Dulieu, F. Masnou-Seeuws, and P. Pillet, *Phys. Rev. Letters*, 1998, 40, 4402–4405.
20. http://www.webelements.com/ .
21. See, e.g., H. Eyring, J. Walter, and G. E. Kimball, *Quantum Chemistry*, 1944, J. Wiley & Sons, New York; chapter 7, pp. 92–99.
22. W. M. Haynes, Ed., *CRC Handbook of Chemistry and Physics*, 92nd edition, 2011, CRC Press, Boca Raton.
23. D. A. Garland, and D. M. Lindsay, *J. Chem. Phys.*, 1984, 80, 4761–4766.
24. W. Kleinig, V.O. Nesterenko, P.-G. Reinhard, and Ll. Serra, *Eur. Phys. J. D*, 1998, 4, 343–352.
25. E. Wigner, and F. Seitz, *Phys. Rev.*, 1933, 43, 804–810.
26. R. Cortés, A. Tejeda, J. Lobo, C. Didiot, B. Kierren, D. Malterre, E. G. Michel, and A. Mascaraque, *Phys. Rev. Letters*, 2006, 96, 126103 1–4.

8 Electron Transport in Metals

A metal is a lot more than a lattice of atoms in their electronic ground states. In the atomic lattice state, for many metals the only bond formation would be due to van der Waals-induced dipole interactions. Those interactions are hardly up to the task of stabilizing a solid. Bonding is a matter of electronic overlap between singly occupied or vacant atomic functions to form the delocalized one-electron bonds that characterize the valence level of metals. These massively degenerate wave functions arise from singly occupied atomic functions that combine to form densely packed (in the energy coordinate) bands of molecular wave functions. In atoms where there are no unpaired electrons at the valence level, one or more electrons must be promoted* to a higher energy orbital to permit bond formation. Bonding in metals has been traditionally studied under the title of band theory, where the focus was on delocalization and conduction properties. Here, we will make an effort to introduce both the bonding and the electron delocalization aspects of the problem.

BAND THEORY IN NORMAL COORDINATES

In Chapter 7 we pointed out the differences in chemical bond formation for metals and insulators. The differences are substantial. In insulators, an energy gap forms between the highest occupied orbital and the lowest vacant orbital of the closed-shell molecule. No energy gap is formed between basis wave functions when metals bond and form delocalized one-electron metallic bonds. If the atomic bonding orbitals are half occupied prior to bond formation in metals, they are also half occupied subsequent to bond formation. If you bring a Cs atom up to a large Cs cluster, the bonds that form between the atom and the cluster are one-electron bonds. In the molecular state, the degenerate bonding orbitals are shifted to lower energies. In order to more fully discuss electron transport in metals, we need to develop this idea further, so we can utilize the bonding structure in metals to aid us in understanding electron transport.

We will not be using reciprocal coordinates [1] in the formulation and discussion of band theory of metals. Reciprocal coordinates have been the hallmark of condensed matter physics for more than half a century. A change to ordinary coordinates requires significant reasons to justify the change. If we adopt reciprocal coordinates as the reference frame for condensed matter studies, the price is substantial. In reciprocal coordinates, there is no electron orbital angular momentum quantum number, and no magnetic quantum number in quantum mechanics [2]. If we were

---

* Promotion energy is the energy that must be put into an atom in its electronic ground state to form an open shell valence structure that can form bonds with adjacent atoms.

to continue with use of reciprocal coordinates for MO theory in metals, we should abandon our effort to understand the pattern of superconducting elements at one bar in the periodic table. The orbital angular momentum quantum number and the magnetic quantum number are the foundations for understanding periodic properties of the elements. These two quantum numbers are not available to quantum mechanics in reciprocal coordinates. One of our main objectives is to understand the periodic occurrence of superconductivity at one bar in the elements. We need to keep the tools to do this, at least for the exercise. As a consequence we will forgo advantages that rise from reciprocal coordinates in the hope of finding a solution to the periodic pattern of superconducting elements at one bar in the table of elements, Table 2.1.

Experimental and theoretical data on band structure in solids are available in the form of plots of electron binding energy versus density of states (see Figure 8.1).

Most of the momentum energy scattering data in Figure 8.1 originated in approximately the top 2 nm of the beryllium sample that was mounted as a target in the x-ray energy–momentum scattering reported by Sashin et al. [3]. Figure 8.1 reports results of this thin film sampling of beryllium vapor deposited on carbon film; it is not surprising that the surface topography shown in Figure 8.1 is different from that shown in Figure 8.2, which was generated from bulk data. The overall density-of-states contour in Figure 8.2 was obtained by theoretical calculations using a self-consistent electronic potential [4].

Figures 8.1 and 8.2 present the density of states for beryllium under specific conditions of thickness and temperature. This data represents the degeneracy of states as a function of electron binding energy. Data in Figure 8.2 show the basis functions, s or p, that are associated with a given energy. Determination of the energy in Figure 8.2 where the density of s-basis wave functions goes to zero is a challenge that has not yet been completely dealt with. Changes in temperature will change the relative populations of specific occupied states; however, detailed experimental studies regarding these changes have not yet been found in the current literature.

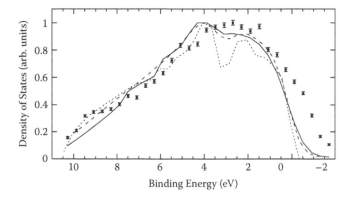

FIGURE 8.1 Density of occupied conduction band states for beryllium from electron momentum spectroscopy profiles (from V. A. Sashin et al., *J. Phys.: Condens. Matter*, 2001, *13*, 4203–19), points and error bars from experiment, lines from Monte Carlo (solid) and linear muffin tin orbital calculations, with two different instrumental convolutions, normalized to unity with energy relative to the Fermi level.

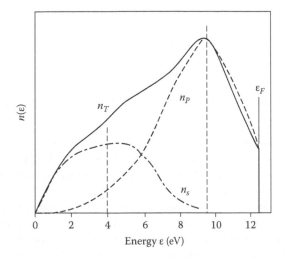

FIGURE 8.2 Density of states for clean beryllium (from R. G. Musket, and R. J. Fortner, *Phys. Rev. Letters*, 1971, *26(2)*, 80–2). n_T, calculated electronic density of states n_p, partial density of states for electrons with p symmetry, obtained from x–ray emission studies [7].

Data presented in Figures 8.1 and 8.2 are very revealing about the band structure in beryllium and, by inference, other metals. Beryllium has only two valence-level electrons. At the valence level, the electronic ground state for atomic Be is $[He]2s^2$. This atomic configuration is not compatible with any bonding for beryllium, so one electron must be promoted to a $2p$ orbital to form bonds in the solid. By looking at Figure 8.2, you can surmise that the valence-level electronic configuration in beryllium is $2s^12p^1$ at the temperature that was used for collection of the n_p data. Bonding in metal conduction bands involves delocalized one electron bonds. So, the block of electronic states shown in the figure represents a large body of mobile, unpaired electrons. Electron mobility in bands suggested in the figure is a function of electron energy relative to the Fermi level in the metal. More tightly bound electrons have lower mobilities.

RESISTIVITY/CONDUCTANCE

ELECTRON SPIN IN OPEN CIRCUIT CONDUCTORS

In a metal system with singly occupied molecular orbitals in conduction bands, it is possible to have an open electrical circuit between a source of electricity and a ground. To accomplish the transfer of electrons from the source to the ground, electrons are added to the system at the source end and, simultaneously, electrons are removed from the system at the ground end, leaving the singly occupied system orbital unchanged. These circuits have unpaired electron spins. Electrons are the particles that move to complete the circuit. Nano-gap junctions between otherwise continuous conductors are crossed by electron tunneling.

In electrical conduction with spin-paired electrons, the only experimentally documented circuits are closed circuits. The electrical circuit in benzene, C_6H_6, is a

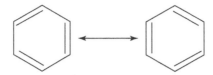

FIGURE 8.3 Valence bond structures for benzene, C_6H_6, a spin paired, closed circuit conductor.

closed zero-electron spin electrical circuit that supports a well-studied ring current. Imagine the motion of the electron pair π-bonds, suggested by the two drawings in Figure 8.3. All of the benzene orbitals are doubly occupied by electrons,[*] so there are no unpaired electrons in the system.

It is possible to utilize benzene-like structures, for example, carbon nanotubes or graphene, in open circuits that conduct electricity. In these cases, the systems demonstrated thus far have always had unpaired-electron carriers. In graphite- or graphene-like conductors, it is possible to establish an unpaired electron current using the vacant carbon wave functions in the system (see the discussion of conduction in an experiment with bilayer graphene in Chapter 11).

It is possible to have unpaired-electron current carriers in organic systems containing π-electron structures if the systems are doped so that the π-electron conductors are radical cations or radical anions. It is also essential to have appropriately charged, nonconductor, counter ions in the matrix to balance the charges of the organic radical ions (see typical structures in Figure 8.4).

At the end of his career, R. B. Woodward was deeply involved in a project to construct an organic chemical superconductor [8]. The schemes that he produced included a number of graphene-like organic sheet π-electron systems. These molecules would be capable of spin-paired closed circuit conduction in a single plane. Many of them are also likely to be capable of being superconductors with appropriate doping to produce unpaired electron conductors. This is the means by which the currently known organic superconductors operate.

Organic chemical conductors and superconductors have developed into a mature branch of condensed matter physics and chemistry [9,10]. Figure 8.4 illustrates the chemical structures of four organic compounds with crystals that are conductors; under the right conditions of temperature and pressure they are also superconductors.

Drawings like those in Figure 8.4 can carry almost zero information content for individuals not expert in organic chemistry or at least trained in the interpretation of organic structural drawings. These drawings are here to provide a sense of size and shape for the conductor stacks within the various crystal structures that manifest the conductivity of these materials. In all four organic structures in Figure 8.4 the carbon–carbon bond lengths can be approximated by the bond in benzene, 140 pm, 1.4 Å. The thickness of the π-electron structure above and below the plane of each ring is of

---

[*] In the simplified valence drawings in Figure 8.3, carbon is represented by vertexes, the six hydrogens are only represented by unfulfilled valence on carbon. Valence for carbon is four. There should be four electron pair bonds at each vertex, which represents the location of a carbon atom. The missing valence is due to the convention that valence bonds from carbon to hydrogen are not drawn, and the hydrogen atoms are understood to fulfill all of the unspecified valences.

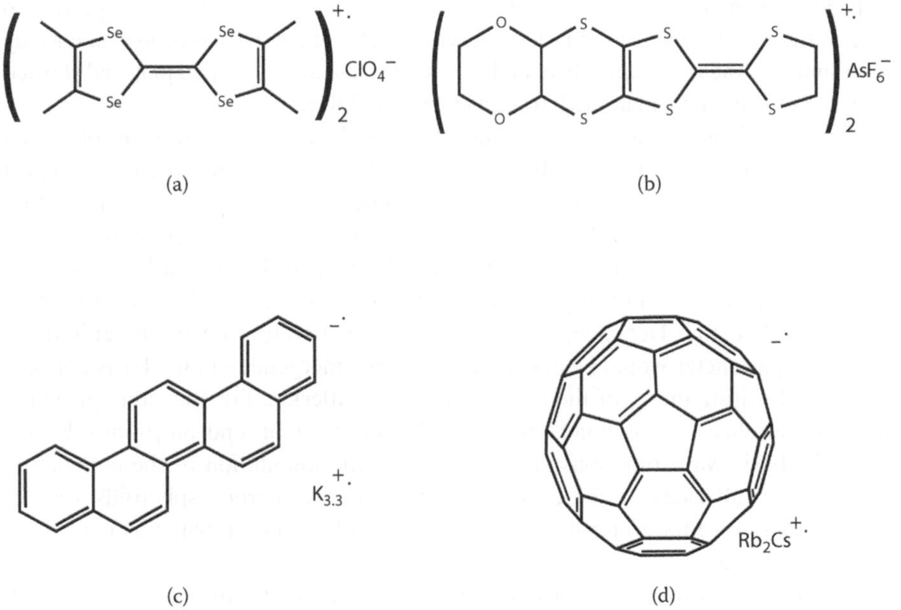

(a)   (b)

(c)   (d)

FIGURE 8.4 Examples of organic superconductors: (a) di–(tetramethyl–tetraselena–ful-valene) perchlorate, (TMTSF)$_2$ClO$_4$ (from K. Bechgaard et al., *Phys. Rev. Letters*, 1981, *46*, 852–5); (b) (1,4–dioxane–2,3–diyldithio)dihydro–tetrathiafulvalene, DODHT (from H. Nishikawa et al., *J. Amer. Chem. Soc.*, 2002, *124*, 730–1); (c) K$_{3.3}$–piceneide (from R. Mitsuhashi et al., *Nature*, 2010, *464*, 76–9); (d) Rb$_2$Cs–fulleride (from R. M. Fleming et al., *Nature* 1991, *353*, 787–8).

the order of 70 pm. These molecules function as conductors in a crystalline matrix by electron conduction through crystallographically stacked π-electron systems.

All four of the examples of organic conductors in Figure 8.4 are odd-electron π-delocalized electronic systems. The first one bar organic superconductor was reported by Klaus Bechgaard and coworkers in 1981, di-(tetramethyl-tetraselena-fulvalene)-perchlorate, Figure 8.4a [11]. Earlier reports of organic chemical super-conductivity came from the same group using elevated pressures to overcome the Mott insulator phase transitions for the first examples, at temperatures above the expected superconducting critical temperature [12]. The thia- (S) and selana- (Se) fulvalene organic structures like that shown in Figure 8.4a are often referred to as Bechgaard salts. A recent Bechgaard salt superconductor is shown in Figure 8.4b [13]. This unsymmetrical, bulky donor* requires external pressure, 16.5 kbar, to attain a transition to superconductivity at 3.3 K.

Picene is a polynuclear aromatic hydrocarbon that is the basis for the organic super-conductor illustrated in Figure 8.4c [14]. In this application, the level of potassium doping is critical to the superconducting critical temperature. The solid illustrated in Figure 8.4c had a doping level of K$_{3.3}$ per picene, and a one-bar critical temperature of

---

* The two six-membered rings in Figure 8.4b are *cis* fused, which significantly alters the stacking pat-tern for the π-electron systems. See the illustration of the x-ray structure in the original paper [13].

18 K. Decreasing the potassium doping level to 2.9 reduced the critical temperature to 7 K. Of the alkali metals in group 1, evidence of superconductivity was present for only potassium, K, and rubidium, Rb, doped picene; the cesium, Cs, doped material showed evidence of a metal insulator (Mott) transition near 150 K [14].

The last of the organic superconductors we will mention is an example of an alkali metal-doped buckminsterfullerene, C_{60} Rb_2Cs [16], Figure 8.4d. The report by Fleming et al., included a plot of fulleride superconducting critical temperature as a function of the crystal lattice parameter, a_0 in Å. Of the compounds included in the study, C_{60} Rb_2Cs had the highest T_c at 31.5 K, and the largest lattice parameter, ~14.5 Å. The relationship between the variables was, to the eye, nonlinear with limited scatter. Dependence of the critical temperature on the magnitude of the lattice parameter would be consistent with the emergence of an s-basis conduction band limiting the T_c of the superconducting fulleride. For the same principal quantum number, s-basis bonds are generally shorter than end-on p-basis bonds. End-on p-basis wave functions are known to be the foundation of the conducting bands in the fullerides. In the conducting state, the two carbon spheroids are sufficiently close together to form an end-on carbon p-basis bond, which is part of the conducting band.

The orbitals involved in conduction for all of the examples in Figure 8.4 are well understood from a molecular orbital point of view. There are detailed *ab initio* calculations available in the literature for all of the structures in Figure 8.4. There are calculations of conducting band structure based on the organic molecular orbitals.

Formation of the doubly charged ions in the solid state for the organic structures in Figure 8.4 has not been reported. Formation of the doubly charged ions in the solid state is unlikely on energetic grounds. The neutral organic structures corresponding to those shown in the figure are all formally insulators.* If doubly charged ions are involved, the size variance of the organic structures in Figure 8.4 is such that the distance between paired electrons would differ substantially if the charges for one pair were spread over two monomers.

Existence of this family of superconductors offers an opportunity to experimentally establish the spin multiplicity of charge carriers in organic superconductors. Experiments probing the carrier spin multiplicity could be conducted in either the quantum Hall regime or a Josephson junction. In our searches of the literature we have not been able to locate reports of definitive experiments establishing the spin multiplicity of the current carriers in superconductors, with the exception of strontium ruthenate, Sr_2RuO_4. This material has recently been shown to support half quantum vortices in the superconducting structure. The half quantum vortices have been definitively connected to triplet carriers in the superconductor.†

Experiments that pin down the spin states of carriers in organic superconductors would not be easy experiments. Nonetheless, thin films of any of the conductors in Figure 8.4 should respond to spin probes as varied as magnetic circular dichroism or studies of the Knight shift under conditions of zero electron scattering.

---

* Thin films of picene with atmospheric oxygen have been explored for use as an organic field effect transistor [15]. The presence of oxygen is essential for the semiconductor properties.
† See the discussion of electron spin in superconductors, Chapter 4.

The nonintuitive features of the impact of electron spin or zero spin on conductance in simple circuits are easily matched by the nonintuitive features of resistivity of elemental conductors, which is our next topic.

Resistivity Phenomenology

Resistivity, ρ (Ω·m), is a bulk property of conductors. It is the product of resistance, R (Ω), and area of the conductor, A (m²), divided by its length, L.

$$\rho = R\frac{A}{L} \tag{8.1}$$

Resistivity values are temperature/pressure dependent and are reported at zero current. How are we going to explain the pattern shown in Figure 8.5?

Looking in the literature, you will find little effort to account for the pattern shown in Figure 8.5 or Table 8.1. Metals shown in Table 8.1 differ in resistivity by two powers of 10 for Mn and Ag; a factor of 100 is big for both chemists and physicists. We have found no discussion of the cause, at least for the massive gap between manganese and silver.

Figure 8.5 suggests that the primary source of the pattern shown is what has been called *promotion energy* in metals. Promotion energy is the energy that must be added to a collection of atoms in their ground state to produce a bulk metal that can conduct electricity. Conduction bands for metals are all at the valence level, which means that the bands are generally composed of linear combinations of *ns* and *np* orbitals, where *n* is the maximum occupied principal quantum number in the ground

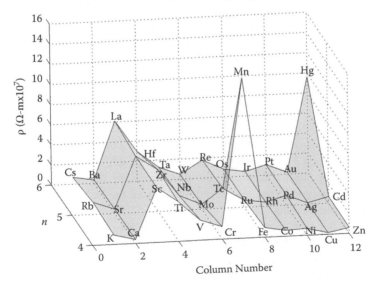

FIGURE 8.5 Bulk resistivity at ambient conditions for main group and transition metals as a function of *n* and column number in the periodic table.

TABLE 8.1

Resistivity of Metals (*10^{-7} $\Omega \cdot$m) at ~293 K

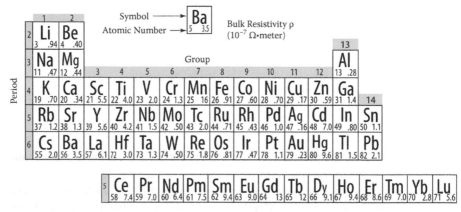

state of the atomic metal. Some of the irregularities in the figure can be seen as non-linear trends in elemental groups and are removed by plotting the data on the basis of resistivity in units of Ω-atom, where atom is the atomic diameter of the atom in the crystal. The major irregularities in the table are for mercury, Hg, and manganese, Mn; these remain to be accounted for. Resistivity for gadolinium, Gd, is also a spike in an extended plot.* It is surely significant that Gd occupies the same location in the *f* transition series as Mn does in the *d* transition series.

Increase in resistivity at the start of the transition metal series, group 3, is also prominent and not given to a simple explanation. The valence level conduction bands for column 3 metals are nominally the same as those for column 2 elements. Higher electron promotion energies are certainly involved in the relatively increased resistivity seen in column 3. The literature reports on the carrier density of the group 3 metals include those by C. Reale [17] and Kevane et al. [18]. The early measurements by Kevane et al. [18] employed the Hall effect and gave a value of 2.9 carriers per atom for both yttrium and lanthanum. The measured room temperature carrier densities for scandium, yttrium, and lanthanum were reported by Reale as: Sc, 0.41, Y, 0.45, and La, 0.48 electron/atom [17]. The large difference between these measurements of carrier density for lanthanum are at this point unexplained. The structural basis for group 3 high resistivity conceivably could arise from the promotion energy required to attain a minimally conducting ns^2np^1 electronic configuration for these metals. This conductor configuration could be the basis for the robust superconductivity of lanthanum.

As the number of *d* electrons increases in the interior of the metal, the resistivity more or less systematically decreases, (see columns 4, 5, 6, Figure 8.5 and Table 8.1). Group 7, just halfway through the *d* transition series, is always an exception in both chemistry and physics. Half-filled subshells of electrons are unusually stable,

---

* You can visualize the spike for Gd by looking at the data in the bottom line of Table 8.1. Gd has the highest resistivity in the line. It is only a factor of 2 or 3 in this case, but it is still a spike.

nonbonding, and nonconducting. An identical pattern is observed in the resistivity of the lanthanides, which reaches a maximum when the f subshell is half filled at Gd (see Table 8.1). Clear explanations for the influence of inner-shell electrons on the resistivity of elemental metals are not yet available. It seems likely that the magnetic effects of a half-filled inner subshell that is spin aligned on the mobility of valence level unpaired electrons is substantial. This magnetic localization effect could play a significant role in the bond strength for manganese. Similar effects should operate in the refractory metals. Sorting out the real sources of binding energy in, for example, refractory metals will not be a simple process simply because of the astronomical number of magnetically active particles involved. Development of new tools for the study of this subject within the context of existing molecular orbital approaches [19,20] may be a necessary next step.

Column 11 in the periodic table, the copper family, has the smallest resistivity of any of the columns of elemental metals. Silver, Ag, in group 11, is the most efficient elemental conductor. During World War II, a large calutron magnet was wound with silver wire for separation of uranium radio nuclides in the Manhattan Project. Many attempts have been made to induce members of the copper family to become superconducting. None of these attempts have been reported as successful to date. Lack of superconductivity in these elements can be explained by their $[X]ns^1$* Fermi level which will always have electron scattering resistivity at the lowest temperatures because of Fermi contact.

Manganese and mercury are both qualitatively different in resistivity from the other metal elements in Figure 8.5. Gadolinium has more than half the resistivity of manganese, and a greater resistivity than mercury. Both gadolinium and manganese have half-filled inner subshells in their atomic ground states. In the same column of the periodic table as manganese, technetium and rhenium have roughly a factor of 8 to 9 lower resistivity than manganese. A firm basis for understanding these facts has not yet been developed. It seems likely that magnetic effects of unpaired electrons will be involved. Self-consistent field molecular orbital theory should, in principle, be capable of ultimately providing insight into these questions [19,20], though novel treatments of one-electron chemical bonds may be required for reliable convergence of the calculations.

CAUSES OF ELECTRON SCATTERING IN CONDENSED METALS

It has been known since the late 1900s that resistivity in metals arises from electron scattering. At low temperatures, electrons in the highest energy conduction band follow the Fermi surface and will scatter from there. Electron scattering from the Fermi surface associated with an atom in the lattice can be quantum mechanically described by partial wave scattering [21].

Historically, electron scattering in metals has been treated as a direct analog of classical Bragg x-ray scattering in metallic crystals. The origins of this analogy are well documented [22]. X-radiation for crystallography has the archetypical

---

* X in brackets represents the electron shell configuration of Ar, Kr, and Xe, respectively, for Cu, Ag, and Au.

wavelength of 0.154 nm for Cu K_α radiation. de Broglie wavelength for a thermal electron at ambient 300 K is given by $h(2\pi m_e k_B T)^{-0.5}$,* or 4.305 nm, a factor of 28 larger wavelength than copper K_α radiation. If we drop to 3 K in temperature, the ratio goes up by a factor of 10 to approximately 280. Bragg scattering is still the model and phonons are the scattering agents at temperatures in the 3 K range.

Electron scattering in atoms and metals is quantum mechanically controlled, as required by the Sommerfeld equation (Chapter 6) by a process that has been known since the early 1930s: partial wave scattering [21]. This is the process that describes scattering of an electron from the Fermi surface of an atom in a lattice. It is the process that we need to use to capture the atomic quantum mechanical influences on both electron scattering and superconductivity.

MOLECULAR ORBITAL APPROACHES TO RESISTIVITY

Ohm's law, Equation (8.2), tell us that the potential drop, V, along a circuit is proportional to the current I, with the proportionality being R, the resistance.

$$V = IR \tag{8.2}$$

There is also a mechanism in electrical resistance that permits thermalization of nonthermal electron velocities. This part of the relationship is dealt with in the Sommerfeld relationship, Equation (6.1). For modern physics and chemistry, the problem is how to realize this picture within the framework of quantum mechanics.

$$\frac{\kappa}{\sigma T} = \frac{\pi^2}{3}\left(\frac{k_B}{e}\right)^2 \tag{8.3}$$

Sommerfeld's relationship and the quantization of electronic phenomena in metals, for example, Equation (6.4), the von Klitzing constant, restrict our choices of a model for resistivity. Sommerfeld's relationship is the connection between electrical conductivity, σ, thermal conductivity, κ, temperature, T, and three fundamental constants, π, e, and k_B. Boltzmann's constant, k_B, enters the relationship because of the Boltzmann distribution of particle velocities in a thermal equilibrium. Electrical conductivity enters the relationship through use of electronic wave functions of an electron gas. Vibrational wave functions of lattices were not considered in the development of the relationship [23]. A foundation for conversion of electron kinetic energy to kinetic energy of atomic nuclei was discovered and reported by Enrico Fermi in 1930 [24,25] during his studies of hyperfine structure in atomic spectroscopy. The atomic spectroscopy application was devoted to the exchange of nuclear spin angular momentum between an atomic nucleus and an s-wave function electron in an atom. If the electron has a probability of occupying the same geometric coordinates as the nucleus of the atom, as happens in atomic s basis functions and Fermi

---

* h is Planck's constant. m_e is the mass of the electron. k_B is Boltzmann's constant. T is temperature in kelvin.

contact, during the period of contact, it is possible for the electron and nucleus to exchange spin angular momentum, linear momentum, and kinetic energy. When the geometric coordinates of the two particles are identical, the electron is formally in the interior of the nucleus.

Participation of electron scattering in resistivity is very well known and well documented [26,27]. Fermi contact has been widely studied in the conducting electrons in metals with respect to the electron nuclear contact component of the Knight shift, (see Chapter 4). If Fermi contact happens in the conducting electrons of metals, these same electrons will experience partial wave scattering when their velocity distribution differs from the thermal distribution of electron velocities. Velocity distributions for conducting electrons are changed from the thermal distribution whenever an electric potential, V, or a magnetic field, B, is introduced. The potential will make manifest both conductance and resistivity in the metallic media. For magnetoresistance resulting from introduction of a magnetic field, see the discussion of the Hall effect and other topics in Chapter 9.

Approaches to the treatment of partial wave scattering in atomic and molecular systems are not at all new to quantum mechanics. The general form for the radial wave equation for the case of a single hydrogen-like atom [28].

$$P_{n,i}^2(r) = \frac{(n-l-1)!Z}{n^2\left[(n+1)!\right]^3}\left(\frac{2Zr}{n}\right)^{2(l+1)}\Psi^2(r) \tag{8.4}$$

where the electronic wave function term, $\Psi^2(r)$, approaches a constant value as r approaches 0.

The partial wave equations for Coulomb scattering are well known, and are given good coverage in most texts for quantum mechanics, (see, for example, Ref. 21). The wave mechanical solutions for Coulomb scattering are tedious because of the infinite range of the Coulomb potential. For scattering inside metals, a full Coulomb potential is not realistic, as Gauss' law requires that the electric field be zero external to a closed neutral electron carrier. The use of a screened Coulomb potential in this case is much more realistic for obtaining scattering amplitude in a metal.

A hard sphere potential, Equation (8.5), can be solved analytically (see Shankar [29]) and exhibits a threshold behavior. Equation 8.6 is true for a wide variety of interactions, including a screened Coulomb potential. In Equation (8.5), r_0 is the range of the scattering potential, a value that must be explicitly calculated, determined by experiment, or estimated.

$$V(r) = \infty \quad r < r_0 \tag{8.5}$$
$$= 0 \quad r > r_0$$

Partial wave scattering phase shifts have been derived without approximation by Shankar, and they exhibit the threshold relations [30] as the wave vector approaches zero,

$$k \to 0 \tag{8.6}$$

$$\tan \delta_l = \delta_l \propto (kr_0)^{(2l+1)}$$

In this equation, and throughout this book, k is the wave vector.* δ_l, is the phase shift, which depends on l, the electron orbital angular momentum. Shankar points out [31], "The above $(kr_0)^{2l+1}$ dependence of δ_l at low energies is true for any reasonable potential, with r_0 being some length scale characterizing the range." If kr_0 is small, then so is δ_l.

The phase shift is important, because it provides information essential to the scattering amplitude, the feature that has our interest. In quantum mechanics, the scattering amplitude is the amplitude of the scattered, spherical wave, that is generated in a scattering interaction involving a planar incoming electron wave. This is one step closer to the scattering cross-section, the feature that we are looking for.

Sakurai gives the partial scattering amplitude, f_l, [32] as

$$f_l = \frac{e^{i\delta_l} \sin \delta_l}{k} \tag{8.7}$$

From Equation (8.7), and Sakurai's expression for the full scattering amplitude, $f(\theta)$, can be rewritten as [33]

$$f(\theta) = \frac{1}{k} \sum_{l=0} (2l+1) \left(\frac{e^{2i\delta l} - 1}{2ik} \right) P_l(\cos \theta) \tag{8.8}$$

$$= \frac{1}{k} \sum_{l=0} (2l+1) e^{ei\delta l} \sin \delta_l P_l(\cos \theta)$$

$$= \sum_{l=0} (2l+1) f_l P_l(\cos \theta)$$

If kr_0 is small, so is δ_l small. Equation (8.7) can be rewritten as

$$f_l \frac{(\cos \delta_l + i \sin \delta_l) \sin \delta_l}{k} = \frac{\delta_l}{k} + O(\delta_l^2) \tag{8.9}$$

Using the threshold value for the l dependent phase shift, Equation (8.6), we arrive at an expression that can be evaluated,

---

* An electron wave vector is the vector whose magnitude is $2\pi/\lambda_e$. Wave vector magnitude is referred to as wave number. λ_e is the electron wavelength.

$$f_l \cong \frac{\delta_l}{k} \propto r_0 (kr_0)^{2l} \qquad (8.10)$$

Sakurai gives the total scattering cross section as [34]

$$\sigma_{tot} = \int |f(\theta)|^2 \, d\Omega \qquad (8.11)$$

$$= \frac{4\pi}{k^2} \sum_l (2l+1) \sin^2 \delta_l$$

If the cross section from the partial wave l is defined as σ_l, and the right side of Equation (8.11) rewritten, we obtain

$$\sigma_{tot} = \sum_l \sigma_l = 4\pi \sum_l (2l+1) |f_l|^2 \qquad (8.12)$$

This leads directly to

$$\sigma_l = 4\pi(2l+1) |f_l|^2 \qquad (8.13)$$

$$\cong 4\pi(2l+1) \left(\frac{\delta_l}{k} \right)^2$$

$$\cong 4\pi(2l+1) r_0^2 (kr_0)^{4l}$$

where the final, threshold result uses details from both Equations (8.6) and (8.10). Equation (8.13) is controlling for partial wave scattering cross sections at threshold energies. It is not convenient to deal with this equation, because it contains r_0 and k. We have no reliable route for converting the cross section to actual electron scattering, or resistivity because the scattering length and the wave vector at threshold are not available with useful accuracy. At a minimum, the partial wave cross section-to-electron scattering conversion would involve setting an empirical parameter for either the partial waves or the full scattering amplitude. To side step these difficulties we will look at partial wave cross sections (scattering) for values of $l > 0$, relative to the corresponding s-orbital partial-wave cross section (scattering).

$$k = \frac{\sqrt{3k_\beta T m_e}}{\hbar} \qquad (8.14)$$

$$\frac{\sigma_{l>0}}{\sigma_{l=0}} = (2l+1)(kr_0)^{4l}$$

Equipartition of thermal energy is a fundamental theorem of statistical mechanics. It states that the average kinetic energy of electrons is $\frac{1}{2}k_B T$ for each degree of

TABLE 8.2

Order of Magnitude Resistivity Ratios versus T

Ratio T	300 K	200 K	100 K	10 K	1 K
p/s	$3*10^{-4}$	$1*10^{-4}$	$3*10^{-5}$	$3*10^{-7}$	$3*10^{-9}$
d/s	$5*10^{-8}$	$1*10^{-8}$	$7*10^{-10}$	$7*10^{-14}$	$7*10^{-18}$
f/s	$8*10^{-12}$	$7*10^{-13}$	$1*10^{-14}$	$1*10^{-20}$	$1*10^{-26}$

freedom, in a solid there are three degrees of freedom. This is the source of the 3 in the top equation in Equation (8.14).

In Table 8.2 we show the estimates at this level of approximation for relative electron scattering, resistivity of p-, d-, and f-basis conduction bands as compared to corresponding s-basis conduction bands, at temperatures that range from ambient, ~300 K, to very low temperature, ~1 K. First the average electron wave number was calculated as a function of temperature using a Boltzmann distribution, giving a value for the average magnitude of the wave vector, k, of a thermal distribution of electrons, in wave number, m^{-1}. For the value of r_0 in this exercise, we chose 100 pm, 1 Å. This is roughly two-thirds of the diameter of a d-series transition metal.[*] It is also approximately twice the Bohr radius—the radius of a Bohr hydrogen $1s$ orbital. Relative values in Table 8.2 depend upon r_0; however, for values not far from 1 Å, the qualitative structure of the table is not changed. With these numbers, the temperatures and electron angular momentum quantum numbers associated with Table 8.2, construction of the table was readily automated in a spreadsheet.

The table shows that for equal populations of conduction bands, most of the resistivity would be carried by s bands. The number of atomic orbitals generated in the buildup (aufbau) of atomic structure depends explicitly on the atomic quantum numbers of the system. There is only one s orbital for every principal quantum number. When $l = 0$, s orbitals, the value of the magnetic quantum number, m, is also zero. When $l = 1$, as in p orbitals, there are three possible values for m, –1, 0, and 1. This produces three p orbitals for every value of n. When $l = 2$, there are five d orbitals corresponding to the five allowed values for m, –2, –1, 0, 1, and 2. Inner subshell electrons, d and f electrons, are not often involved in electrical conduction because the current is generally carried by electrons that are at the surface of the atom, that is, at the valence level. In high-temperature superconductors where the current is carried in copper $3d$-basis conduction bands, the copper is present as the doubly charged cation, Cu^{2+}. The valence shell electrons of copper have been removed in this case by ion formation.

Table 8.2 suggests that at sufficiently low temperatures, electron scattering in p-, d-, or f-basis conduction bands can become small enough that it could not be distinguished from zero scattering or resistivity. That is the condition of superconductivity. At this point, the question of other sources of electron scattering is not resolved. The Sommerfeld relationship points to electronic wave functions as the dominant source

---

[*] The lattice parameter for the copper cubic close-packed lattice is 361 pm, 3.61 Å. The lattice parameter for this lattice is the distance between unit cells, and for this lattice it corresponds to the diameter of a copper atom in the lattice.

of electron scattering and thermal equilibration of the lattice. The real values for the Lorenz number for metals show that there are contributions to resistivity from other sources, but they are small. Partial wave scattering is the only electron-scattering mechanism that is consistent with the Sommerfeld relationship.

Molecular orbital (MO) modeling of electron transport in metals is not new. It began with studies by Wigner and Seitz in 1933 [35]. With many variations it has continued to the present. Quantized electrical conductance studies require some molecular orbital theory for development of an understanding of the process in any particular instance [36]. Yaliraki et al. utilized an *ab initio* formalism with scattering based on a generalized Landauer expression [37]. A sampling of recent papers applying molecular orbital theory to electron transport produced papers that dealt with single molecule electron transport [38–42], charge transfer band structure [43], and quantum transport [44]. Molecular and quantum electron transport are both areas that require a modern molecular orbital approach. If partial wave scattering is included in studies of resistivity in metals, these studies as well will require a full molecular orbital treatment.

Partial wave scattering of electrons has been part of the core curriculum in atomic and molecular quantum mechanics since the details of the subject were worked out in the mid-twentieth century [21]. It has not been studied as part of the curriculum in condensed matter physics because it was not invented when the reigning paradigm for resistivity was being established at the close of the first third of the twentieth century. Partial wave scattering is a well known and robust feature of molecular quantum mechanics that has the features needed for the development of an exclusively electronic theory of conductance and resistivity, as required by both the Sommerfeld relationship and the quantum Hall effect.

Partial wave scattering in a metal conduction band will quantitatively account for scattering of electrons as a function of temperature and electron orbital angular momentum associated with the specific conduction wave function. This formalism also provides the direct connection between electrical conductivity, (S•m^{-1}), and thermal conductivity, (W•m^{-1}•K^{-1}), required by the Sommerfeld relationship.

ELECTRON PHONON SCATTERING

Quantum mechanics insures that electron phonon scattering is always a possibility when electrons are moving in the vicinity of phonons. A substantial fraction of the scattering phenomena that have been explained in the literature as electron scattering from phonons may be the result of partial wave scattering from nuclei [21], generated exclusively in the electronic wave functions of the system. In partial wave scattering, there is independent experimental evidence to confirm the reality of the phenomenon going back to the original spectroscopic experiments of Fermi [24]. As far as we have been able to determine, there are no reports in the literature that establish which mechanism is actually involved in cases where partial wave scattering from nuclei can compete with electron-phonon scattering. This situation includes a very large fraction of the reported cases of electron scattering in metals.

Independent evidence of electron caused phonon scattering exists in a number of places in the literature [45]. Phonon lasers are presently available to a number of investigators [46]. It would be possible with a phonon laser to arrange a cross-beam scattering experiment with an electron beam *in vacuo* to provide information about the effectiveness of the phonon scattering of electrons. Definitely, the cross sections for electrons scattering from phonons and phonons scattering from electrons are the same. The question is: "Can we demonstrate electron scattering from phonons by measuring electron scattering as a function of phonon energy and number density?" Search of the literature for the results of such an experiment has not yet been fruitful. The difficulty is the substantial difference in the masses and momenta of conducting electrons and phonons. A single collision of a phonon with an electron can produce an observable result. In the case of electrons scattering from phonons, it is probable that multiple events will be required to get above the detection threshold. Partial wave scattering of electrons has been demonstrated unequivocally many times. The same is not true for electrons scattering from phonons in our search of the literature.

The Sommerfeld relationship suggests that the electron scattering processes, central to conductance and resistivity, arise from an electronic wave function. Partial wave scattering of electrons offers an electronic scattering mechanism known to occur in real samples. Partial wave scattering provides a route for energy equilibration between conducting electrons and the lattice of nuclei. We have not found another mechanism that meets the requirements established by the success of the Sommerfeld relationship between thermal and electrical conductivity [23].

ELECTRON CYCLOTRON RESONANCE IN METALS

Equation (8.15) is the fundamental equation for electron cyclotron resonance in a low-density ionic plasma in a vacuum.

$$m_e \omega_c = e B_\square \tag{8.15}$$

When the cyclotron resonance experiment is done in a metal sample, Equation (8.15) does not balance when the measured cyclotron frequency, ω_c, the two electron constants, m_e and e, and the applied magnetic induction, B_0, are used. The standard approach to this problem has been to introduce the electron effective mass, m_e^*, which is used to make the measured cyclotron frequency, magnetic induction, and electron charge balance.

$$\omega_c = \frac{e B_0}{m_e^*} \tag{8.16}$$

The successes of self-consistent field MO calculations of the properties of metals and junctions suggests that, in quantum mechanics, the effective mass of the electron in metals is unchanged from its normal value. The imbalance of Equation (8.15), when it is used for electron cyclotron resonance in metals is not due to an effective change in mass of the electron, but due to an unexpected change in the momentum of the electron in a metal matrix. The cause of the observed electron effective mass appears to be magnetoresistance. This subject will be discussed in some detail in the next two chapters.

Anisotropic magnetoresistance was first observed by William Thomson (Lord Kelvin) in an 1856 experiment. This is the anisotropy of resistance to current flow as a function of geometry for a ferromagnet. Magnetoresistance, as it is commonly known, was first observed in the Hall effect, discovered about 23 years later. Discussion of the Hall effect opens the next chapter.

REFERENCES

1. C. Kittle, *Introduction to Solid State Physics, Eighth Edition,* 2005, Wiley New York. See Chapter 2, Wave diffraction and the reciprocal lattice.
2. U. Mizutani, *Introduction to the Electron Theory of Metals,* 2001, Cambridge University Press, Cambridge.
3. V. A. Sashin, M. A. Bolorizadeh, A. S. Kheifets, and M. J. Ford, *J. Phys.: Condens. Matter,* 2001, *13,* 4203–19.
4. T. L. Loucks, and P. H. Cutler, *Phys. Rev.,* 1964, *133,* A819–29.
5. R. G. Musket, and R. J. Fortner, *Phys. Rev. Letters,* 1971, *26(2),* 80–2.
6. T. L. Loucks, and P. H. Cutler, *Phys. Rev.,* 1964, *133,* A819–29.
7. T. Sagawa, *Soft X-Ray Spectrometry and the Band Structure of Metals and Alloys,* D. J. Fabian, Ed., 1968, Academic Press, New York; p. 29.
8. M. P. Cava, M.V. Lakshmikantham, R. Hoffmann, and R. M. Williams, *Tetrahedron,* 2011, *67(36),* 6771–97.
9. R. J. Thorn, *Chemical Equilibria Bases for Oxide and Organic Superconductors,* 1996, Wiley, New York.
10. A. G. Lebed, *The Physics of Organic Superconductors and Conductors,* 2008, Springer, New York.
11. K. Bechgaard, K. Carneiro, M. Olsen, F. B. Rasmussen, and C.S. Jacobsen, *Phys. Rev. Letters,* 1981, *46,* 852–5.
12. D. Jerome, A. Mazaud, M. Hibault, and K. Bechgaard, *J. Phys. (Paris), Lett.* 1980, *41,* L95.
13. H. Nishikawa, T. Morimoto, T. Kodama, I. Ikemoto, K. Kikuchi, J.-I. Yamada, H. Yoshino, and K. Murata, *J. Amer. Chem. Soc.,* 2002, *124,* 730–1.
14. R. Mitsuhashi, Y. Suzuki, Y. Yamanari, H. Mitamura1, T. Kambe, N. Ikeda, H. Okamoto, A. Fujiwara, M. Yamaji, N. Kawasaki, Y. Maniwa, and Y. Kubozono, *Nature,* 2010, *464,* 76–9.
15. H. Okamoto, N. Kawasaki, Y. Kaji, Y. Kubozono, A. Fujiwara, and M. Yamaji, *J. Amer. Chem. Soc.,* 2008, *130,* 10470–1.
16. R. M. Fleming, A. P. Ramirez, M. J. Rosseinsky, D. W. Murphy, R. C. Haddon, S. M. Zahurak, and A. V. Makhija, *Nature* 1991, *353,* 787–8.
17. (a) C. Reale, *Appl. Phys. A,* 1973, *2(4),* 183–5; (b) C. Reale, *Phys. Stat. Sol.,* 1973, *58,* K5.
18. C. J. Kevane, S. Legvold, and F. H. Spedding, *Phys. Rev.,* 1953, *91,* 1372–9.
19. See, e.g., W. J. Hehre, L. Radom, P. v.R. Schleyer, and J. A. Pople, *Ab Initio Molecular Orbital Theory,* 1986, John Wiley & Sons, New York; or reference 15.
20. R. G. Parr, and W. Yang, *Density-Functional Theory of Atoms and Molecules,* 1989, Oxford University Press, New York.
21. J.J. Sakurai, *Modern Quantum Mechanics,* revised ed., S.F. Tuan, Ed., 1994, Addison, Wesley, Longman, New York, 7.6 Method of Partial Waves, pp. 399–409.
22. F. Bloch, *Proc. R. Soc. London A,* 1980, *371,* 24–7.
23. A. Sommerfeld, *Die Naturwissenschaften,* 1927, *15,* 825–32.
24. E. Fermi, *Nature,* 1930, *125,* 16–7.
25. E. Fermi, *Collected Papers,* Vol. I, 1962, University of Chicago Press, Chicago.
26. See, e.g.: N.W. Ashcroft, and N.D. Mermin, *Solid State Physics,* 1976, Thompson Learning; p. 315; and Ref. 21.

27. M. Marder, *Condensed Matter Physics*, second edition, 2010, J. Wiley, New York; Chapter 16, p. 453 .

28. J.C. Slater, Quantum Theory of Matter, second edition, 1968, McGraw-Hill, New York; p. 121, Equation (6–26).

29. R. Shankar, *Principles of Quantum Mechanics*, second edition., 1994, Springer, New York; p. 549.

30. *Ibid.* p. 550, equation 19.5.29.

31. *Ibid.* p. 550.

32. J.J. Sakurai, *Modern Quantum Mechanics*, revised edition, S.F. Tuan, Ed., 1994, Addison Wesley Longman, New York, p. 402, Equation (7.6.17).

33. *Ibid.* p. 402. Sakurai's Equation (7.6.17) includes the first two lines in Equation (8.8). The third line is just a formal contraction of the line above.

34. *Ibid.* p. 403, Equation (7.6.18).

35. E. Wigner, and F. Seitz, *Phys. Rev.*, 1933, *43*, 804–10.

36. See, e.g., S. N. Yaliraki, A. E. Roitberg, C. Gonzalez, V. Mujica, and M. A. Ratner, *J. Chem. Phys.*, 1999, *111(15)*, 6997–7002.

37. R. Landauer, *Philos. Mag.*, 1970, *21*, 863.

38. P, A. Derosa, and J, M. Seminario, *J. Phys. Chem. B*, 2001, *105*, 471–81.

39. Y. Xue, S. Datta, and M. A. Ratner, *J. Chem. Phys.*, 2001, *115(9)*, 4292–9.

40. J. Taylor, H. Guo, and J. Wang, *Phys. Rev. B*, 2001, *63*, 245407 1–13.

41. J. B. Neaton, M. S. Hybertsen, and S. G. Louie, *Phys. Rev. Letters*, 2006, *97*, 216405 1–4.

42. S. M. Lindsay, and M. A. Ratner, *Adv. Materials*, 2007, *19(1)*, 23–31.

43. Y. Xue, S. Datta, and M. A. Ratner, *J. Chem. Phys.*, 2001, *115(9)*, 4292–9.

44. J. Taylor, H. Guo, and J. Wang, *Phys. Rev. B*, 2001, *63*, 245407 1–13.

45. See, e.g., S. Lehwald, J. M. Szeftel, H. Ibach, T. S. Rahman, and D. L. Mills, *Phys. Rev. Letters*, 1983, *50*, 518–21.

46. See, e.g., B. S. Williams, H. Callebaut, S. Kumar, Q. Hu, and J. L. Reno, *Appl. Phys. Letters*, 2003, *82(7)*, 1015–7.

47. M. Marder, *Condensed Matter Physics*, second edition, 2010, John Wiley & Sons, New York; Section 16.2.3 Effective Mass, Equation (16.28), p. 459.

9 Magnetoresistance

Magnetoresistance is the resistance to motion of electrons caused by the presence of an external magnetic field. The Hall effect [1] tells us that magnetoresistance happens. Magnetoresistance is the increase in resistance in an electric circuit* when the circuit is exposed to an external magnetic field. It is maximum when the current in the circuit is perpendicular to the applied field and is approximately zero when the current is parallel to the applied magnetic field.

Voltage regulators often depend upon magnetoresistance. Consumers are not generally aware of the existence of voltage regulators, let alone their mode of operation. Magnetoresistance is a manifestation of electron scattering resistance that was introduced in Chapter 8. Like electrical resistance, this topic is central to understanding the foundations of superconductivity, and it will be discussed here. The most general manifestation of magnetoresistance is known as the Hall effect.

HALL EFFECT

Magnetoresistance was first explored in metals by Edwin Hall's 1879 discovery of the development of a voltage perpendicular to a current flowing in a gold foil perpendicular to a magnetic field†—the Hall effect [1]. Roughly 40 years lapsed between Hall's experiments and Kapitza's exploration of the increase in resistivity associated with placing a conductor in a perpendicular orientation to a magnetic field [2]. Both the Hall effect and magnetoresistance depend upon cyclotron motion of electrons in a magnetic field that is perpendicular to an electron current. In the Hall effect, cyclotron motion of the electrons in the current creates a classical electron momentum path that is cycloidal. This path results in a charge bias from one side of the conductor foil to the other that creates the Hall voltage, V_H. In pure metals, for example, gold, the Hall voltage generated in Figure 9.1, is given by

$$V_H = \frac{-IB}{ned} \tag{9.1}$$

Figure 9.2 is a drawing of a standard Hall bar.‡ Voltage measured between connection posts 1 and 3 is the standard Hall voltage, V_H. Posts 1 and 3 in Figure 9.2b are also used for measurement of the Hall conductivity, σ_{xy} and Hall resistance, ρ_{xy}. In Equation (9.1),

---

* Nonferromagnetic conductors are essential for ordinary magnetoresistance. Ferromagnetic conductors produce anisotropic magnetoresistance. The major surface difference between the two is the magnitude of the anisotropy of the magnetoresistance.
† At the time of his discovery, Hall was a graduate student at Johns Hopkins University. He later became a professor there.
‡ A Hall bar, illustrated in Figure 9.2b, is the experimental tool used in Hall effect measurements for a given material.

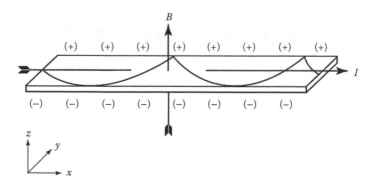

FIGURE 9.1   Classical schematic of electron trajectory in the Hall effect for a pure metal; current and fields follow physics conventions.

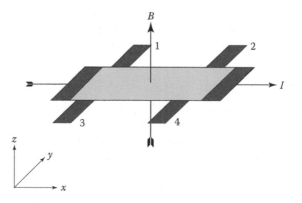

FIGURE 9.2   Standard electrical connection post configuration on a Hall bar.

I is the applied current, *B* is the magnetic flux density, *n* is the carrier density for the metal of the foil, *e* is the electron charge, and *d* is the thickness of the foil.

It is possible to see from Figure 9.1 that the classical path of an electron traveling under the influence of a driving voltage on the *x* axis will be in the area of the Hall bar longer if the magnetic field, *B*, is present than if there were no field. This path length effect was thought to be the source of magnetoresistance studied by Kapitza [2].* The Hall effect is the development of a voltage in the *y* axis when there is a current flowing in the *x* axis of the magnetoresistance experiment illustrated in Figure 9.1. Hall effect measurements of magnetic field strength were, for a long period, the only reliable method for obtaining magnetic field data. There is a large supporting literature both scientific and commercial. Studies of pure magnetoresistance in polycrystalline metal samples have not, in comparison received much attention, though a small number of papers on the angular anisotropy of magnetoresistance are very informative.

---

* P. I. Kapitza went on from his studies of magnetoresistance to discover superfluidity in helium-4, ⁴He, in 1937. He received the Nobel Prize in 1978 for superfluidity, 50 years after his investigation of magnetoresistance.

HALL COEFFICIENT

Equation (9.2) gives the definition of the Hall coefficient. It is the ratio of the applied electric field on the x-axis to the product of the magnetic flux density times the Hall current density on the y-axis, j_y.

$$R_H = -\frac{E_x}{Bj_y} \tag{9.2}$$

R_H is the conventional label for the Hall coefficient. ρ_{xy} conventionally designates Hall resistivity.

Positive Hall coefficients occur significantly less often than negative Hall coefficients. Materials with positive Hall coefficients are found in a few elemental metals and p-type semiconductors. Generally, materials with positive Hall coefficients are not classed as good conductors. For example, the semiconductor, diamond lattice tin,[*] has a positive Hall coefficient. This crystalline form of tin is never involved in superconductivity. Allotropic phase transitions between diamond lattice tin and tetragonal lattice tin are examples of "near-Mott transitions." The sparsely populated semiconductor band gap in diamond lattice tin is close to pushing tin to an insulator phase, and as a consequence diamond lattice tin has zero properties as a superconductor, other than its atomic scale identity with superconducting tetragonal, distorted-face centered cubic lattice tin.

An explanation of the cause for the negative electron current density, j_y, that is responsible for positive Hall coefficients is presented in the treatment of Landau levels in Chapter 10. Arguments presented there indicate that negative electron orbital angular momentum states, which exclusively populate the lowest Landau level in Hall experiments, are responsible for observation of a positive Hall coefficient.

ANGULAR ANISOTROPY OF MAGNETORESISTANCE

In 1960, Alekseevskii and Gaidukhov published a study of the angular anisotropy of magnetoresistance in single crystals of elemental metals [3]. This was one of the first, and remains one of the most informative, manuscripts on the subject of magnetoresistance in elemental metals at low temperatures. Alekseevskii and Gaidukhov's results for magnetoresistance in silver metal are shown in Figure 9.3. The scientific culture at that time did not include inclusion of precise detailed experimental descriptions in papers discussing scientific findings, so at least some of the details of the experiments have been left as an exercise for the reader. Figure 9.3 was described as having been gathered with the "crystal [001] axis approximately (±5°) along the specimen axis" [4]. From the context in the manuscript, we presume that the magnetic field was aligned with the [001] crystal axis using backscattering Laue photography. Marder has evidently reached a similar conclusion, at least for the silver experiment in Figure 9.3 [5].

---

[*] The low-temperature form has a diamond lattice. It is also referred to as α-tin or grey tin The form that is stable at temperatures above 13.2° C has a tetragonal lattice. This form is also called β-tin, white tin, or metallic tin The high-temperature form also has the highest density.

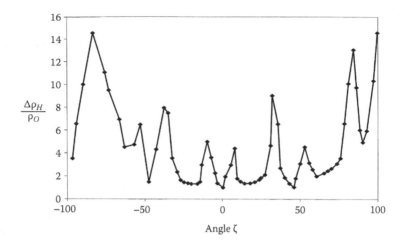

FIGURE 9.3 Magnetoresistance anisotropy of single crystal silver at 4.2 K. (From N. E. Alekseevskii, and Y. P. Gaidukhov, *Soviet Phys., J. Exper. Theor. Phys.*, 1960, *17*, 481–4. FIG. 2 caption, p. 482.)

Within the author-estimated experimental accuracy of the angle measurement, Figure 9.3 shows local minima in the relative resistance curve at angles of ± 0, 45, and 90 degrees. Around each local minimum there are local peaks in resistivity, and between the local peaks there are somewhat elevated local minima in resistivity with diffuse inflection points in the range of ± 20–25 and 65–70 degrees. It is worthy of note that the relative resistance in both the local minima and maxima increase non-linearly between 0 and 90°. The pattern seen in Figure 9.3 can be explained using the hypothetical experimental sketch shown in Figure 9.4.

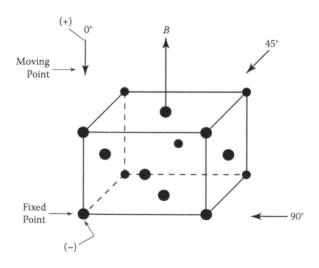

FIGURE 9.4 An experimental schematic sketch, consistent with the results reported by Alekseevskii and Gaidukhov (see Figure 9.3).

Figure 9.4 illustrates a sketch of what may have been the experimental set-up used by Alekseevskii and Gaidukhov, in which individual atoms in a one unit cell of the single crystal of silver metal they used are represented by black balls. The model is a face-centered cubic unit cell structure. The magnetic field is aligned with one of the cubic axes (since they are all the same, it is an arbitrary choice). Resistance measurement was accomplished with an electrical circuit; (+) and (−) poles are illustrated with an orientation parallel to the magnetic field, $\zeta = 0°$ (ζ is the angle in Figure 9.3).

If you look at Figures 9.3 and 9.4 you can see that the local minima at $\zeta = 0°$, 45°, and 90° are associated with the relative resistance measurement being made in a plane of silver atoms of the crystal. When the measurement is away from the plane of atoms on either side, there is an increase in resistance because you have moved away from the optimum overlap of the planar conducting path. Why is it that the peaks in relative resistance on either side of a conducting plane of atoms fall off to what are only slightly elevated local minima? An explanation mined from the original paper is moderately complex. In the same paper Alekseevskii and Gaidukhov present data collected using a single crystal of tin, shown in Figure 9.5.

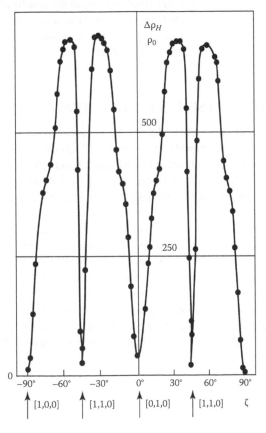

FIGURE 9.5 Tin, single crystal, magnetoresistance anisotropy at 4.3 K. (From N. E. Alekseevskii, and Y. P. Gaidukhov, *Soviet Phys., J. Exper. Theor. Phys.*, 1960, *17*, 481–4; FIG. 8 a, p. 483.)

Experimental results in Figure 9.5 appear to be generally compatible with the schematic diagram in Figure 9.4, with the exception of the axis of rotation for the relative resistance sampling circuit. Figure 9.4 shows the axis of rotation to be a crystal axis that is perpendicular to the magnetic field axis. Data in Figure 9.5 suggest that the axis of rotation for the resistance sampling circuit was coincident with the magnetic field axis, as there is no apparent increase in magnetoresistance with the angle ζ in Figure 9.5 as there is in Figure 9.3. Changes in relative resistance with ζ in Figure 9.5 can be reasonably attributed to changes in atomic orbital overlap as a function of ζ in the molecular wave functions that make up the conduction bands in tin at 4.2 K. Effective conversion of the four peaks in Figure 9.5, tin, to four valleys in silver, Figure 9.3, is probably a cyclotron motion effect reflecting different orientations of the magnetic field and the electric current in the two experiments. As the angle ζ was changed in Figure 9.3 from 0° to 90° (see Figure 9.4), the angle between the magnetic field and the plane of travel of the electrical current changes from 0°, no Hall effect, to 90°, maximum Hall effect.[*] There is no surprise that the relative resistance and the value of ρ_o both change and by different amounts.

One of the clear lessons that comes from looking at the results of the Alekseevskii and Gaidukhov experiments is that the space in which current-carrying electrons moves in metals is highly structured. Electronic structure in conductors has been part of this field at least since introduction of the Fermi surface near the middle of the last century [7].

CONDUCTION BAND SHIFTS WITH EXTERNAL MAGNETIC FIELDS

Magnetic field interactions with conduction band electrons have been studied extensively in the de Haas van Alphen effect. Magnetic effects on conduction band energies have been observed in spin splitting of conduction bands [8], exchange interactions [9], and temperature-dependent exchange splitting [10]. Many of the papers in this general area are a consequence of the high interest that exists in giant magnetoresistance and related topics.

Well-known and documented Zeeman effects [11] in atomic spectroscopy provide the atomic theoretical basis for anticipating magnetic field effects on the energies of conduction band electrons in metals. At low temperatures, where searches for superconductivity are often focused, these effects should be most easily noticeable. Conduction bands that are linear combinations of atomic basis functions with $l > 0$ will have L, molecular orbital angular momentum, > 0, and will be susceptible to relative energy changes caused by external magnetic fields. The original Stern–Gerlach experiment [12] demonstrated magnetic separation of different atomic electronic states in 1922.

---

[*] Using Hall-effect terminology presumes a thin foil carrier of the current, which is not the case here. The magnetoresistance will be more complex than the ordinary Hall setup.

MEISSNER–OCHSENFELD EFFECT

Meissner and Ochsenfeld published a brief report in 1933 on their experiments with superconductors and external magnetic fields [13]. They were following reports from the de Haas laboratory in Leiden on the steepness of the transition curves as a function of the measuring current for tin and other superconductors [14]. What they found was:

> . . . our experiments on tin and lead have given the following results:
>
> 1. In cooling below the transition point the distribution of lines of force in the region exterior to the superconductor is altered and becomes essentially as expected for a body of zero permeability, or of diamagnetic susceptibility [per unit volume]
>
> $$\frac{-1}{4\pi}.^{*}$$
>
> 2. In the interior of a long lead tube, in spite of the change of magnetic field in the exterior as described under 1, in descending below the transition point the previously established magnetic field is essentially undiminished [13–15].

Discovery of zero magnetic permeability for a bulk superconductor is arguably the most important discovery in the study of the Meissner–Ochsenfeld effect. When magnetic permeability is zero in SI units, volume magnetic susceptibility is –1 and external magnetic fields, with field strengths below the critical field, H_c, are excluded from the bulk material.

Magnetic susceptibility, χ, is, in general, a tensor, and is defined by the partial derivative of the magnetization, M, with reference to the magnetizing field, H as

$$\chi = \frac{\delta M}{\delta H} \qquad (9.3)$$

For linear magnetic response,[†] Equation (9.4) gives the relationship between magnetic susceptibility, χ_v, and magnetic permeability, μ, in SI units.

$$\chi_v = \frac{\mu}{\mu_0} = -1 \qquad (9.4)$$

Although χ is dimensionless in SI and cgs/Gaussian units, there is a conversion factor of 4π in changing to SI units from cgs units. Meissner and Ochsenfeld reported the permeability of bulk superconductors as 0 and the susceptibility as $-1/4\pi$ (dimensionless) in cgs units [13]. Superconductor volume susceptibility in SI units is –1 (dimensionless).

Magnetic permeability, μ, is the relationship between magnetic induction, B, and magnetic field, H, in a linear magnetic medium. In practice, a linear magnetic medium refers to a lack of ferromagnets in the system.

---

[*] Gaussian units.
[†] In practical terms, this means no ferromagnets.

$$B = \mu H \tag{9.5}$$

To facilitate understanding of the Meissner–Ochsenfeld effect, it will be worthwhile to review the sources of diamagnetism and paramagnetism in elemental metals.

DIAMAGNETIC AND PARAMAGNETIC ELEMENTAL METALS

Diamagnets have a negative magnetic susceptibility that can be experimentally observed. Magnetic levitation of pyrolytic carbon is an example. In pyrolytic carbon the levitation utilizes the π conduction bands of graphite in the sample, which generate a magnetic field that opposes the externally applied field. The induced magnetic field floats the wafer of carbon above the perturbing magnet. At this writing, at least two toys are on the market that utilize this technology. It is the use of Landau diamagnetism of the π-electrons in graphite that is the source of the diamagnetism.

The most commonly encountered diamagnetic materials are organic molecules. They are insulators and, on average, lack conduction bands. Organic molecules like cellulose from wood, cotton, or other plant sources, cannot generate a repulsive magnetic field large enough to levitate even a very thin wafer of the material. Diamagnetic materials include water, copper, silver, mercury, lead, bismuth, and many other items of commerce.

Diamagnetism for the elements in the periodic table can be understood in terms of Landau diamagnetism [16,17], which arises ultimately from the negative charge on the electron. Paramagnetism is a consequence of electron spin effects of which there are several. Paramagnets can be thought of as weakened ferromagnets. In ferromagnets, electron spin(s) in one unit cell are quantum mechanically aligned with spin(s) in adjacent cells throughout the magnetic domain.

Recent values for molar magnetic susceptibilities in SI units for elemental metals are listed in Table 9.1. Values for four metals are blank in Table 9.1. These are the ferromagnetic elements (see the following subsection of this chapter). Lanthanide elements that directly follow gadolinium, Gd (that is, terbium, Tb; dysprosium, Dy; holmium, Ho; erbium, Er; and thulium, Tm), are the most paramagnetic, nonradioactive elements in the periodic table. This attests to both the significance of a half-filled f subshell to paramagnetism, as well as the importance of these elements to magnet technology in the modern economy.

Every field of science is plagued by some background level of urban legend. In condensed-matter physics and chemistry, is the idea one of the persistent oldies that "diamagnetism requires an absence of unpaired electron spins in the material." Table 9.1 shows diamagnetic elements in columns 1, 2, 11, 12, 13, and 14. Those columns contain metallic elements. Metallic materials have the order of 10^{23} unpaired electrons per mL. A complete list of columns that included diamagnetic elements would include columns 15 through 18.

If you look carefully at Table 9.1, you will see that one requirement for a diamagnetic metal is that the metal not have an open d or f subshell. All of the diamagnetic metals in the table fulfill this requirement. The requirement is met in columns 1, 2, 11, 12, 13, and 14. There are nine metals, shown in Table 9.1 in those seven columns,

TABLE 9.1
Molar Magnetic Susceptibility of Elemental Metals

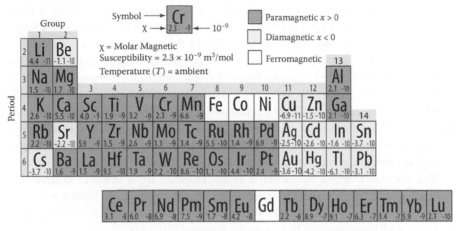

Note: β-tin, Sn, the superconducting crystalline form of the element is paramagnetic; see [20].
Source: http://periodictable.com/Properties/A/MolarMagneticSusceptibility.v.wt.html

that are paramagnetic so a closed or nonexistent inner shell, d or f, does not guarantee a diamagnetic metal. Purity of the elemental sample is somewhat more important for measurements of magnetic susceptibility than it is for many other measurements. When the paramagnetic molar susceptibility is in the range of 10^{-10} m^3/mol, an ultra-trace of a ferromagnetic material can produce that much paramagnetism.

For the elements at the border of diamagnetism and paramagnetism—for example, the elements in group 14 of the periodic table—relatively small changes can induce a shift from diamagnetic to paramagnetic susceptibility. The tin magnetic susceptibility quoted in Table 9.1 corresponds to that of diamond lattice tin, also known as α-tin. This is the grey tin form that is stable below 13.2° C. It is a semiconductor. The metallic form of tin, β-tin, is the tetragonal form of tin. This form of tin is paramagnetic. Its volume paramagnetic susceptibility has been the subject of at least two reports in the literature [19,20]. Hoge [20] first reported the magnetic anisotropy of tetragonal β-tin crystals. He gave the values of the volume magnetic susceptibility perpendicular and parallel to the tetragonal tin crystal axis. Translated into SI molar magnetic susceptibilities, comparable to Table 9.1, those values are $\chi \perp$, 4.0×10^{-11}; $\chi \|$, 3.6×10^{-11} m^3/mol. The change in magnetic susceptibility for tin at the allotropic phase transition between β- and α-tin is just an elaboration of the consequences of the near-Mott transition that occurs in the α- to β-tin phase transition. The two different forms of tin have different crystal structures, specific resistivities, and also different magnetic permeabilities. The change in the magnetic permeability in this case is explicitly due to the change in chemical bond structure in the two forms of elemental tin. Only β-tin is a superconductor.

Bismuth, Bi, a fascinating semimetal, has the most negative molar diamagnetism in the periodic table at -3.6×10^{-9} m³/mol. Bismuth is in period 6 of group 15, to the right of lead, Pb. The most paramagnetic element in the nonradioactive periodic table is terbium, Tb, at 1.2×10^{-6} m³/mol. Its paramagnetic susceptibility is three orders of magnitude larger than bismuth's diamagnetism. The extremes in a series are often informative. Terbium is in the middle of the lanthanide series of elements; all but the last two members of the series are strongly paramagnetic. It is just to the right of gadolinium, a ferromagnet. Bismuth is adjacent to the bottom right edge of the periodic table. Bismuth is at the low end of electron communication for metals. By the simplistic scheme in Table 7.1, bismuth is classed as a metal. The bonds that hold bismuth together are one electron delocalized bonds. Bismuth is not an approximation of an electron spin-paired material, yet it is the most diamagnetic element including carbon. Carbon bonds with electron pair bonds.

The discussion above shows that it is not essential for a material to have zero electron spin to have a magnetic susceptibility that is negative. No known materials, other than active bulk superconductors, have a magnetic susceptibility of -1 (volume, unitless), which corresponds to a magnetic permeability of 0.

Diamagnetism and paramagnetism in metals does not depend directly upon electron spin in the conduction bands of the metals. All of the metals shown in Table 9.1 have unpaired electron spins in their conduction bands. There are no known exceptions to the unpaired electron spin of the conduction bands of normal metals. Differences between paramagnetic and diamagnetic metals must be explained by the response of local electron spin angular momentum to an applied magnetic field,

$$\left(\frac{\partial M}{\partial H}\right)_{\text{Local}} \tag{9.6}$$

and communication of the local spin response within the Condon magnetic domain [21] of the metal, to generate a domain susceptibility,

$$\left(\frac{\partial M}{\partial H}\right)_{\text{Domain}} \tag{9.7}$$

that can contribute to the macroscopic response. Communication of local spin response to external magnetic perturbations is built into the structure of ferromagnetic materials and is well understood. A similar type of local spin response communication to remote parts of the domain appears to be built into Condon domains for diamagnetic and paramagnetic metals, though it is not as yet widely recognized. From analysis of changes in electronic heat capacity at the transition to superconductivity, it appears that the mechanism for domain-wide communication of local spin response to external fields is mediated by electron scattering. See the discussion of this topic later in this chapter.

London suggested the involvement of the exchange interaction in the communication of electron spin [22]. We suggest that electron scattering in the magnetic domain is the dominant mechanism for mediating this susceptibility exchange interaction in metals.

Condon restated the classical expectation of the existence of magnetic domains whenever the differential magnetic susceptibility,

$$\chi = \frac{\delta M}{\delta H},$$

exceeds unity. In the definition of differential susceptibility, the change in magnetic field, δH, is the perturbing influence. Change in the magnetization, δM, is the response of the system. The hidden question is: "How is the response mediated?" Measurements of electronic heat capacity at the critical temperature for the superconductivity phase transition discussed below suggest that electron scattering is the mediator that defines differential magnetic susceptibility in nonferromagnetic metals.

There is no doubt that spin pairing will produce diamagnetic materials such as cellulose or other fully bonded organic molecules. There is also no doubt that archetypical metals like copper can have a volume diamagnetic susceptibility. The position differences in the periodic table between the diamagnetic and paramagnetic metals suggest that spin-orbit electron interactions in the conduction bands may be the key to the difference between paramagnetism and diamagnetism for ordinary metals. The pattern of diamagnetic elements strongly suggests a role for electron spin, m_s, (s in chemistry), and orbital angular momentum, l, in mediating the paramagnetic coupling of conduction band electrons in paramagnetic metals. Scattered distribution of diamagnetism in columns 1 and 2* suggests that the quantum mechanics of diamagnetism will not be given to simplistic analysis.

Elemental Ferromagnetism and Paramagnetism

There are four ferromagnetic elements in the periodic table: iron, Fe; cobalt, Co; nickel, Ni; and gadolinium, Gd. Magnetic susceptibility is not defined for these elements below the Curie point, because their response to an external magnetic field depends upon the magnetic history of the sample—magnetic hysteresis. Ferromagnetic materials have a large, variable positive magnetic susceptibility and exhibit hysteresis. These materials form magnetic domains in which the magnetic moments of individual atoms in the domain are aligned. Communication of electron spin alignment within a single magnetic domain appears to occur through the intermediacy of electrons that carry electronic specific heat in the conduction band. Evidence supporting this observation can be seen in plots of electronic specific heat as a function of temperature in the region of the Curie temperature[†] for a ferromagnet (see Figure 9.6 [23,24]). The sharp points in Figure 9.6 are at the Curie temperatures.

---

* In the alkali metals and alkaline earths, the diamagnetic elements are cesium, Cs, in column 1, and beryllium, Be, and strontium, Sr, in column 2, (see Table 9.1).

† Curie temperature is the temperature above which a ferromagnet becomes a paramagnet, which has a linear dependence of magnetic moment on applied magnetic field strength for small fields.

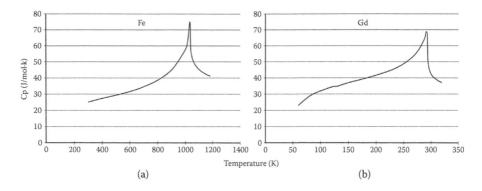

FIGURE 9.6 (a) Electronic specific heat of iron (from L. S. Darken, and R. P. Smith, *Ind. Eng. Chem.*, 1951, *43*, 1815–20), peak maximum, 1033 K. (b) Electronic specific heat of gadolinium (from M. Griffel, R. K. Skochdopole, and F. H. Spedding, *Phys. Rev.* 1954, *93*, 657–61), peak maximum, 292 K. For both curves in Figure 9.6 the maxima are at the Curie temperatures, the locus of the second-order phase transition.

We have not presented the literature graphic for the classically estimated magnetic specific heat for these two metals [25]. It appears that there was a typographical error in the value for the Curie temperature for iron, in the paper by Hofmann et al. [25]. This value (1043 K) has been very widely quoted but does not conform with the maximum in the curve shown in Figure 9.6a, 1033 K [23]. Experimental data that was the source of the curve in Figure 9.6a is presented in the paper by Darken and Smith [23]. None of the classical heat capacity estimates in the paper by Hoffmann et al. could have moved the peak in the curve by 10 K [25]. The Curie point quoted by Hoffmann et al. for gadolinium [25] is congruent with the curve from the heat capacity data, Figure 9.6 b.

Existence of an electronic heat capacity local maximum at the Curie temperature is a feature of the electronic landscape of ferromagnets that is as heuristically important as the local minimum in electronic heat capacity for conductors at the superconducting critical temperature.* This local maximum in heat capacity is a thermodynamic consequence of the second-order phase transition that occurs there between the ferromagnet and its paramagnetic counterpart. Hofmann, et al. discussed the heat capacity local maximum in terms of magnetic effects [25]. Thermal carriers for the magnetic effects must be electrons in the conduction band and/or their surrogates. When we speak of "surrogates," in this case we are referring to the inner subshell magnetic electrons associated with atoms in the lattice. Inner subshell electrons cannot participate in the conduction band because the closed electron subshells above them will drastically decrease the overlap integrals of the inner subshell electrons with their counterparts on adjacent atoms. No overlap means no participation in the conduction band. Thermal electron energy is the second parameter that operates here. When the sum of electronic and thermal energies of the electrons in d or f atomic basis functions reaches the electronic potential energy of the conduction band, the exchange interaction provides a path for exchange of both magnetic and

---

* See Figure 9.8.

thermal energy between the two wave functions. We anticipate that the exchange interaction is the origin of the large increase in the number of carriers of heat capacity near the Curie point. It would be worthwhile to carefully examine the electronic density of states in the four ferromagnetic elements to see if the density of states correlates with the Curie points for these elements. The maxima in the electronic heat capacity at the Curie temperature in Figure 9.6 indicate a sharp increase in the number of electrons that carry heat capacity at the Curie temperature. The only candidates that we have found for these additional heat capacity carriers are the inner subshell d or f electrons in iron or gadolinium. These inner subshell electrons can become heat capacity-carrier surrogates through the exchange interaction with s-basis conduction band electrons, if they are in an appropriate energy state. The exchange interaction between the pairs of electrons on the same atom is contingent upon the energies of the two electrons being in the same narrow range. This information could be seen in a plot of the density of states for the systems.

On the paramagnetic side of the ferromagnetic phase transition, the magnetic susceptibility of the material is a constant associated with the domain. Condon reported an experimental search for domains in his paper on beryllium diamagnetic domains [21]. The dimensions of the Condon domain appear from this report to be not larger than the order of 0.5 mm. This result is consistent with electron scattering in the conduction band being the mechanism for communication of local spin response to a magnetic perturbation to remote parts of the domain.

Electron scattering and the change in heat capacity at the critical temperature in superconductors will be continued in this discussion of the role of conduction band electron scattering in magnetic property communication within metals. Prior to dealing with the heat capacity changes at T_c in superconductors, we will briefly discuss both electron spin and magnetic susceptibility in the superconducting state.

ELECTRON SPIN AND SUPERCONDUCTORS

Knight shifts are observable for superconducting metals only because the Meissner–Ochsenfeld effect requires that a thin surface layer of a superconductor must be permeable to applied magnetic fields below H_c. A Knight shift is the nuclear magnetic resonance, (NMR), chemical shift of an atomic center at the surface of an active superconductor. Since the NMR experiment occurs at the surface, the studies are generally done with films or aggregates of small particles to maximize surface-to-volume ratios. Figure 9.7 presents the Knight shift of the β-allotrope of tin using a linear temperature axis. This plot uses the same data used for Figure 4.1. Linear axis plots have the virtue of presenting a realistic view of the noise in the data. You can see from Figure 9.7 that there is considerably more scatter in the Knight shift data for tin than there is in the critical magnetic field data for the same element (Figure 11.1). Critical temperature for this sample of tin was reported as 3.71 K [26]. The informative feature of Figure 9.7 is that it strongly suggests that the Knight shift for the sample will not be 0 at approximately 0 K. There will be electron spins in the non-superconducting λ layer of the material, and presumably in the bulk of the material itself, at the lowest attainable temperatures.

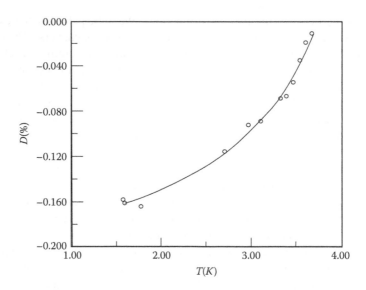

FIGURE 9.7   Knight shift in superconducting tin particles versus temperature on a linear scale. (From G. M. Androes, and W. D. Knight, *Phys. Rev.,* 1961, *121,* 779–87.)

Measurement of the Knight shift under conditions of the quantum Hall regime offers the possibility of directly evaluating the spin state of the conductor under conditions identical to those of a superconductor with the exception of magnetic susceptibility. The available data suggest that this is precisely what happened in the experiments reported by Androes and Knight when the temperature of the measurement was below 3.71 K [26,27]. This is the temperature range shown in Figure 9.7. The experiment was conducted with vapor-deposited tin on nylon. The reported thickness of the tin layer was 40 Å [26,27]. This thickness is too small to support both a top and bottom London penetration layer and a superconducting interior. The London penetration layers would necessarily be smaller than 20 Å thick. Our search of the literature has not found one report of a London penetration layer that was not larger than 20 Å. The drop in the value of the shift, D(%), at the critical temperature of 3.71 K is consistent with a high-order phase transition to the quantum Hall regime. Knight shift data in Figure 9.7 are consistent only with unpaired electrons as the origin. A change in spin pairing at T_c would have appeared as a change in NMR line width, which would certainly have been noticed by the investigators.

It is now known that Sr_2RuO_4 has a superconducting triplet spin state [28]. Earlier studies of the Knight shift in Sr_2RuO_4 also reported a triplet spin state [29]. Based on this evidence and the quantum Hall regime experiment just discussed, it is our considered view that the Knight shift accurately reports the spin state for superconductors that dominantly have unpaired spins in their conduction band.

It would be very useful and informative if a variety of experiments were conducted in the quantum Hall regime to explore the details of the spin states in superconductors.

MAGNETIC SUSCEPTIBILITY, EXPULSION OF EXTERNAL
MAGNETIC FIELDS BY SUPERCONDUCTORS

Meissner and Ochsenfeld's critical observation was recording the expulsion of external magnetic fields from a superconductor at the phase transition critical temperature T [13,14]. They also determined that the magnetic permeability and susceptibility of the bulk superconductor (tubes of β-tin or lead) were, respectively, zero, $\mu = 0$, and $\chi = -1/4\pi$ (Gaussian) or $\chi = -1$ (SI). The London brothers pointed out the probable existence of a limited magnetic field penetration depth, λ, in their classical treatment of the effect [30]. Existence of a thin layer of penetration by an external magnetic field was confirmed by a series of experiments in several laboratories [31]. All of the Knight shift experiments that have been reported with superconductors below T_c depend upon this thin layer of field penetration.

In his paper titled "On the Problem of the Molecular Theory of Superconductivity" [32], Fritz London discussed the phenomenological connection between the quantum mechanical exchange interaction and magnetic susceptibility. A change in magnetic susceptibility, χ, from diamagnetic, $\chi < 0$, or paramagnetic, $\chi > 0$, to what is called pure diamagnetic, $\chi = -1$, is what the Meissner–Ochsenfeld effect is about. Elements that are ferromagnetic below their respective Curie points—iron, cobalt, nickel, and gadolinium—are not superconductors as elements, so we need not be concerned with magnetic induction beyond the linear range. Spin-paired systems, such as cellulose, which is fully bonded and can be obtained with unmeasurably low spin density, typically have negative magnetic susceptibilities [33]; but the susceptibilities are the same general magnitude as the diamagnetism of copper [34], which has unpaired spins in its conduction bands.

The fundamental question is: What is the origin of magnetic susceptibility in metallic materials? For normal metals, the answer appears to be electronic communication of local response to external magnetic fields throughout the Condon magnetic domain of the material. Exchange interactions between electrons [32,35] and dissipative electron scattering are the fundamental mechanism for electronic communication of electron spin information, which provides a basis for magnetic properties in normal metals. These interactions are characterized by the exchange integral [35] in atomic and molecular quantum mechanics. Exchange integrals arise as a consequence of the LCAO MO approximation.

Orbital angular momentum is a second and more subtle basis for magnetic properties of materials. The roles of the magnetic quantum numbers of electrons and the spin quantum numbers of nuclei are progressively more subtle. Relationships that control the total angular momentum of domains of electrons in condensed materials are responsible for generating magnetic properties like susceptibility. At the phase transition to superconductivity there is a switch in magnetic susceptibility from diamagnetic or paramagnetic to -1. The switch is a quantum switch. Magnetic domain size changes from a mesoscale value, determined by the material, to the bulk of the superconductor. A connection between the bulk susceptibility, its value of -1, and the end of dissipative electron scattering will be discussed later (see Figure 9.8 and Chapter 11).

The only systems that are pure diamagnets, $\chi_v = -1$, are bulk superconductors. There is no requirement for electron spin-pairing in solids that have magnetic susceptibility less than zero. Diamagnetic metallic elements are prominently shown in Table 9.1. Metals, elemental or alloys, have unpaired spins in their conduction bands. Fully bonded organic molecules like cellulose are spin-paired and are diamagnetic.* Spin-paired molecules can be prepared with purities high enough that no spins can be detected in the material. Both groups of materials, diamagnetic metals and organic spin-paired molecules, have similar ranges of diamagnetic susceptibilities [33,34].

Meissner and Ochsenfeld's observations of magnetic susceptibility for superconductors at T_c, and the changes in heat capacity that occur at T_c need to be understood as a single package. The following discussion is intended to initiate that understanding.

HEAT CAPACITY AT T_C

Variation of electronic specific heat with temperature for elemental superconductors is one of the historically noted features of superconductivity. Figure 9.8a illustrates a plot of the specific heat of β-tin, Sn, as a function of temperature in the range of ~0.3 to 4.5 K [36]. The figure illustrates the distinct temperature dependence of C_p above and below T_c. Figure 9.8b illustrates the experimental points for reduced heat capacity and temperature for the same data [36] along with the fitted curve for

$$\frac{C_P}{C_{P|T_c}} = a + b \cdot \frac{e^{cT/T_c} + e^{-cT/T_c}}{2} \tag{9.6}$$

where C_P is the superconductor experimental heat capacity, T_c is the critical temperature for the phase transition, and a, b, and c are numeric, least squares fitting constants. For the experimental data of Corak and Satterthwaite illustrated in Figure 9.8a, the fitted parameters for Equation (9.6) are:

a = −0.2119
b = 0.1738
c = 2.558
Chi-square, χ^2 = 0.000204
correlation coefficient square, r^2 = 0.99992538

The small value of Chi-square for the nonlinear regression that produced the fitted curve indicates the closeness of fit for the data produced by Corak and Satterthwaite [36] by Equation (9.6). The data shows real noise, but it may be necessary to consult the original tabular data [36] to verify that. Equation (9.6) is the second equation involving a $\cosh(T/T_c)$ dependence for superconducting tin. The first such

---

* Condon [21] did not consider nonconductors like cellulose. A magnetic domain definition for these materials would not exceed the limits of the molecule and may be limited by electron mobility in molecular structures such as high molecular mass insulators.

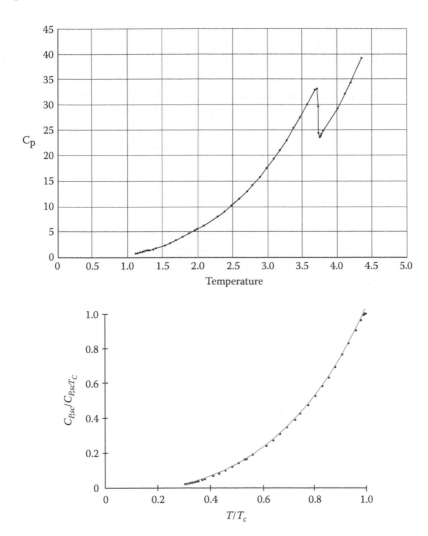

FIGURE 9.8 (a) Specific heat of β-tin, Sn, at low temperature. C_P units, mJ/(mol•K). (b) Experimental points and least squares fitted curve for Equation (9.6). Highest temperature plotted in (b) is 3.7 K. (From W. S. Corak, and C. B. Satterthwaite, *Phys. Rev.,* 1956, *102,* 662–6.)

relationship was described by Andros and Knight in discussion of the Knight shift for superconducting tin [25].

Curves for reduced heat capacity and temperature for the superconducting phase have closely fit the model of Equation (9.6) for all of the superconductors that we have evaluated. Superconductors in this group are:[*] vanadium, V [37]; niobium, Nb [38]; technetium, Tc [39]; indium, In [40,42,43]; and tin, Sn [36,41,42]. Indium has the lowest critical temperature, $T_c = 3.4$ K [43], for this group of elements. We have not found heat capacity data for any superconductor with T_c less than 2 K. Thallium

---

[*] In order of atomic number.

has the lowest critical temperature, 2.4 K [44], for the elements that we have found any heat capacity data for the superconducting phase.

Figure 9.8a shows a significant decrease in the material heat capacity at T_c. We assume that this decrease scales linearly with the number of scattering thermal carriers (electrons), and that the scaling is approximately the same for the super-conductor and the normal metal. For the three available data sets on superconducting tin heat capacity the percentage losses are 31% (Keesom and Koc [41]), 28% (Corak and Satterthwaite [36]), and 30% (Bryant and Keesom [42]):

$$\%scattering = \frac{C_{p,sc} - C_{p,n}}{C_{p,sc}} \cdot 100\% \qquad (9.10)$$

Values for the maximum superconducting heat capacity, $C_{p,sc}$, and the minimum normal metal heat capacity, $C_{p,n}$, were taken from the highest and lowest reported data points surrounding the critical temperature. The distribution of values for the percentage reflects the scatter in the data and variations in the samples. Figure 9.8 presents the data set with the least noise of the three. In normal tin on the high-temperature side of T_c, roughly 30% of the potential carriers of heat capacity are ineffective because of scattering. The scattered electrons have the capacity to communicate local response to external magnetic fields to other parts of the magnetic domain that are within the scattering range. At the low temperature side of the superconducting phase transition, these same electrons lose this capacity to communicate local spin responses because they no longer scatter.

Conduction band electrons that scatter at temperatures barely above T_c must have a minimal level of s-basis wave functions, see Table 8.3. The closeness of the superconducting phase transition insures that conduction band wave functions with basis functions having $l > 0$ will have scattering levels too low to be detected at T_c.

Figure 9.9 presents another of the early Keesom and Kok data sets [44] on elemental heat capacity near the critical temperature for superconductivity. In this figure they present data for the superconducting phase transition for thallium, Tl, in the presence of an external, constant, magnetic field.

Latent heat for the transition shown in Figure 9.9, is the integral of the change in heat capacity for the transition as a function of applied magnetic field from 0 to 33.6 gauss and the range of temperatures for the transition. If the phase transition is referred to as first order, because of the latent heat it is necessary to include the applied magnetic field in the phase.

Thallium is diamagnetic. Niobium, Nb, like β-tin, is an example of a robust elemental superconductor that is paramagnetic as a normal metal. Data for niobium specific heat at the superconductivity transition are shown from a paper by Hirshfeld et al. [45] in Figure 9.10. By 1962 the condensed matter culture had moved away from simple plots of heat capacity versus temperature. In Figure 9.10, heat capacity divided by temperature is plotted against temperature squared to match the approximate temperature dependence of the model for heat capacity that was being used, and to produce the nearly straight lines in the figure.

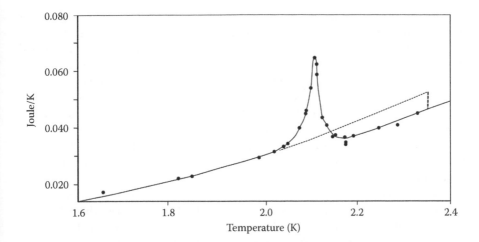

FIGURE 9.9 Heat capacity of thallium as a function of temperature near $T_c = 2.38$ K, in the presence of a 33.6 gauss external magnetic field plotted from the data of Keesom and Kok (from W. H. Keesom, and J. A. Kok, *Physica*, 1934, *1*, 503–12). The dashed lines show the transition in the absence of the applied external magnetic field.

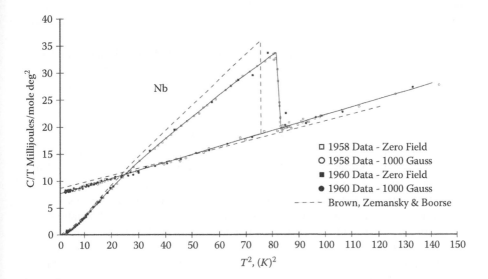

FIGURE 9.10 Heat capacity of niobium divided by T as a function of T^2, near $T_c = 9.25$ K. (From A. T. Hirshfeld, H. A. Leupold, and H. A. Boorse, *Phys. Rev.*, 1962, *127*, 1501–7.) Data for the dashed curve came from the work of Brown et al. [46].

Although the presentation of the data is significantly different between Figures 9.8 and 9.10, by looking at the figures it is possible to generate the notion that the two data sets represent similar behavior for two quite different superconductors: tin and niobium, respectively. Zero-applied-field experiments are shown in Figure 9.10 on the curve that starts approximately at the origin and is discontinuous somewhat above 80 K^2. Niobium critical temperature is 9.25 K [47], or 85.6 K^2. Our estimate for the niobium critical temperature in the experiment illustrated in Figure 9.10 is just over 9 K. Nonsuperconducting heat capacity was obtained with a 1 tesla external magnetic field and is illustrated on the approximately straight line that slops upward from the left in Figure 9.10. The magnetic field is part of the system in the nonsuperconducting state below T_c.

The Born–Oppenheimer approximation allows the separation of nuclear (vibrational) wave functions from the electronic wave functions in cases where the electronic state of the system is more or less independent of the momentum of the nuclei. Most quantum mechanical problems are solved using this approximation. Because of this, it is difficult to remember that superconductivity represents a massively entangled quantum state for which the Born–Oppenheimer approximation does not apply. Motions of the nuclei and the electrons are intimately connected in a superconductor so that separation of nuclear and electronic wave functions may not provide a reasonable approximation for the state of the system The nonseparability of the electronic and nuclear wave functions may be responsible for the cosh functional relationship between the appropriately reduced functions of heat capacity, C_P, and T seen in Equation (9.6). In an active superconductor the number of carriers of heat capacity should include all of the particles in the system. This expansion in the number of carriers of heat capacity could be responsible for the small nonlinearity reflected in the cosh functional relationship between C_P and T. Quantum mechanical relationships utilizing cosh have previously appeared in relativistic quantum mechanics [48], among other places.

The data above suggest that a quantum mechanical theory of superconductivity will be essential for explanation of the details of the heat capacity of superconductors in the framework of a superconductor unified electron–nuclear wave function. Necessity of a quantum mechanical description is even clearer in type II superconductors.[*]

All of the magnetic properties of atoms are involved in the differential magnetic susceptibility of bulk metals. Magnetic alignment of spin states of the nucleus even make small contributions to the resultant magnetization of a metal exposed to a magnetic field. The same is true for electron orbital angular momentum, as well as electron spin angular momentum. Electrons that undergo partial wave scattering from s-basis wave functions have the potential to carry all of this information from the Fermi surface, where they scatter, to the boundaries of the scattering range at T_c. Once scattered electrons reach their destination, they will once again become part of the leading conduction band in the system. This random exchange of conduction band electron position is essential to establishing the electron exchange interaction foundation for magnetic susceptibility in the normal metal Condon domain.

---

[*] See Chapter 12.

The relevant facts concerning electronic interactions and magnetic susceptibility at T_c are:

1. Electron scattering from the conduction band ceases at T_c.
2. No electron spin information is transferred from a formerly scattering Fermi level to distant parts of the magnetic domain.
3. The electron exchange interaction for formerly scattered electrons goes to zero. Exchange interactions for these electrons can still occur, but they will all be local.
4. The bulk magnetic susceptibility of the material goes to –1. This is the extreme value expected for a perfect diamagnet. Electron spin in the conduction band has not changed. The factor leading to a bulk magnetic susceptibility of –1 rise from observations in the Meissner–Ochsenfeld effect and will be discussed in Chapter 11.

Scattering electrons connect the two second-order magnetic phase transitions that we have discussed in this chapter: normal conductor–superconductor transitions at T_c and ferromagnetic-paramagnetic transitions at the Curie point. We know from the many detailed studies of the electronic heat capacity at the critical temperature for superconductors that scattering electrons are explicitly involved on the normal metal side of the superconductor phase transition. In the normal metal, dissipative electron scattering establishes the Condon domain for the metal. The same electrons are implicitly involved on the superconductor side of the phase transition because of the contribution made to the superconductor energy gap by the end of dissipative electron scattering.

The local maximum in heat capacity at the ferromagnetic–paramagnetic transition must be associated with thermally activated electron exchange of the inner-shell unpaired d or f electrons[*] with conduction band electrons that carry electronic heat capacity. The Coulomb exchange interaction has the potential to expand the number of electrons carrying heat capacity without expanding the number of electrons in the conduction band.[†] For the ferromagnetic–paramagnetic transition, the expansion involves the ferromagnetic electrons that are not in the conduction band because of vanishing overlap between inner-shell wave functions on adjacent lattice atoms. The effect of the expansion in the number of heat capacity carriers can be dramatically seen in the heat capacity at the Curie temperature, (Figure 9.6).

It is likely that the de Haas–van Alphen effect will always be a little confusing. This process with all its variations is truly complex. It appears that it is a reflection of the profound complexity of magnetic interactions in systems with astronomical numbers of unpaired electron spins.

---

[*] These are the electrons that control ferromagnetic properties of a metal.
[†] The increase in heat capacity at the Curie temperature was considered in the discussion of ferromagnetism earlier in this chapter.

DE HAAS–VAN ALPHEN EFFECT

In 1930, de Haas and van Alphen reported the discovery of magnetic oscillations as a function of magnetic field in a sample of bismuth, Bi, at 14.2 K [49], Figure 9.11. Shoenberg references Landau's prediction of this effect [50] in Landau's paper on diamagnetism of metals [16,17]. After publication of his paper on the electron gas as a model for metals, Landau worked with Shoenberg in Cambridge. As you might see from his book, Shoenberg was a world expert on the de Haas–van Alphen effect.

Oscillations in magnetization with magnetic field for bismuth were just a tip of the iceberg of the oscillations now referred to as the de Haas–van Alphen, dHvA, effects. When a magnetic field is present, perturbations that are known to induce dHvA oscillations in metals include, in addition to changes in magnetic field, changes in: crystal orientation in a constant field, temperature, and mechanical torque. Shape of the experimental sample is known to be important to the wave forms that are generated.

Data partially illustrating the complexity of the results obtained by thermal- and torque-generated oscillations are shown in Figures 9.12 and 9.13 [21]. Condon's experiments with the de Haas–van Alphen effect resulted in his formulation of a theoretical framework to describe the existence of magnetic domains in diamagnetic and paramagnetic metals. These metals are not magnets by themselves but respond to external magnetic fields to either enhance (paramagnetic) or diminish (diamagnetic) the applied field. Condon's formalism has been adopted by many, and diamagnetic and paramagnetic domains in metals are often referred to as Condon domains.

All of the experiments in the suite of the de Haas–van Alphen effect involve conduction electrons in metals moving in magnetic fields without externally applied electric fields. These are the conditions of electron cyclotron resonance in metals. (See Chapter 6 and Chapter 8.) dHvA conditions are also the conditions of magnetoresistance. In the absence of magnetoresistance there would be no de Haas–

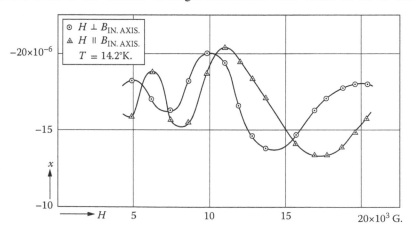

FIGURE 9.11 Oscillations in magnetization, M, as a function of magnetic field, H, in bismuth, Bi, at 14.2 K. (From W. J. de Haas, and P. M. van Alphen, *Proc. Netherlands Roy, Acad. Sci.*, 1930, *33*, 1106; Fig. 6, p. 1116.)

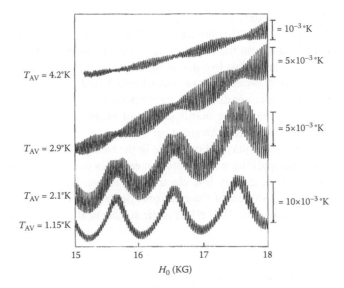

$T_{AV} = 4.2°K$

$T_{AV} = 2.9°K$

$T_{AV} = 2.1°K$

$T_{AV} = 1.15°K$

$= 10^{-3}°K$

$= 5 \times 10^{-3}°K$

$= 5 \times 10^{-3}°K$

$= 10 \times 10^{-3}°K$

H_0 (KG)

FIGURE 9.12 Magneto-thermal oscillations in beryllium, Be, at four different average temperatures. The upward slope of the curves is due to instrumental artifacts. (From J. H. Condon, *Phys. Rev.*, 1966, *145*, 526–35; Fig. 10, p. 532.)

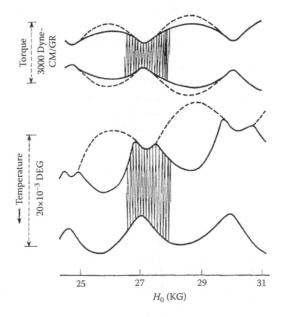

Torque 3000 Dyne-CM/GR

Temperature 20×10⁻³ DEG

H_0 (KG)

FIGURE 9.13 (Upper) Torque and magneto-thermal oscillations of two different samples at different temperatures, same orientation, of magnetic field relative to the crystal and over the same field range [52]. The dashed lines show the behavior expected from two beating periods. (Lower) This figure shows the phase relationship between the envelopes of the magnetization oscillations and the temperature oscillations. The dashed lines show the behavior expected if there were no magnetic feedback effect. (From J. H. Condon, *Phys. Rev.*, 1966, *145*, 526–35; Fig. 11, p. 533.)

van Alphen effect. As a consequence of this, we can anticipate the magnitude of k, the wave vector, in these experiments will reflect the effects of magnetoresistance.

A substantial literature on the de Haas–van Alphen effect in the quantum Hall regime has developed in the last two decades [53]. Proust et al. present high-resolution studies of the dHvA oscillations in underdoped $YBa_2Cu_3O_{6.5}$, a high-temperature superconductor, under conditions of no electron scattering [53]. Conditions of zero electron scattering move the manifestation of magnetoresistance from "magnetic cooling" to quantum magnetoresistance.

The quantum Hall effect plays a larger role in the background of superconductivity than being just the application of quantum mechanics to the Hall effect obtained under conditions of zero, or very low electron scattering. Because of its heuristic importance in the establishment of the electronic description of electrical resistance and conductance, the quantum Hall effect is the subject of the next chapter.

REFERENCES

1. E. H. Hall, *Amer. J. Math.*, 1879, *2*, 287–92.
2. P. L. Kapitza, *Proc. Roy. Soc. A*, 1929, *123*, 292–341.
3. N. E. Alekseevskii, and Y. P. Gaidukhov, *Soviet Phys., J. Exper. Theor. Phys.*, 1960, *17*, 481–4.
4. *Ibid.*, Fig. 2 caption, p. 482.
5. M. Marder, *Condensed Matter Physics*, second edition, 2010, J. Wiley, New York; Figure 17.5 p. 503.
6. N. E. Alekseevskii, and Y. P. Gaidukhov, *Soviet Phys., J. Exper. Theor. Phys.*, 1960, *17*, 481–4; FIG. 8 a, p. 483.
7. L. Onsager. *Phil. Mag. Ser. 7,* 1952, *43*, 1006–8.
8. R. Reifenberger, and D. A. Schwarzkopf, *Phys. Rev. Letters,* 1983, *50*, 907–10.
9. I. A. Merkulov, D. R. Yakovlev, A. Keller, W. Ossau, J. Geurts, A. Waag, G. Landwehr, G. Karczewski, T. Wojtowicz, and J. Kossut, *Phys. Rev. Letters,* 1999, *83*, 1431–4.
10. B. Kim, A. B. Andrews, J. L. Erskine, K. J. Kim, and B. N. Harmon, *Phys. Rev. Letters,* 1992, *68*, 1931–4.
11. P. Zeeman, *Nature*, 1897, *55*, 347.
12. W. Gerlach, and O. Stern, *Z. Phys. A*, 1922, *9*, 353–5.
13. W. Meissner, and R. Ochenfeld, *Naturwissenschaften*, 1933, *23*, 787–8.
14. P. F. Dahl, *Superconductivity*, 1992, American Institute of Physics, New York; Figure 9.3, p. 170.
15. This translation is from, *Ibid.*; p. 178.
16. L. D. Landau, *Z. Phys. A*, 1930, *64*, 629–37.
17. For an English translation, see, D. ter Haar, Ed., *Collected Papers of L. D. Landau*, 1965, Gordon and Beach, London; pp.31–8.
18. http://periodictable.com/Properties/A/MolarMagneticSusceptibility.v.wt.html.
19. K. Honda, and Y. Siuimizu, *Nature*, 1935, *136*, 393.
20. H. J. Hoge, *Phys. Rev.*, 1935, *48*, 615–9.
21. J. H. Condon, *Phys. Rev.*, 1966, *145*, 526–35.
22. F. London, *Phys. Rev.*, 1948, *74(5)*, 562–73.
23. L. S. Darken, and R. P. Smith, *Ind. Eng. Chem.,* 1951, *43*, 1815–20.
24. M. Griffel, R. K. Skochdopole, and F. H. Spedding, *Phys. Rev.* 1954, *93*, 657–61.
25. J. A. Hofmann, A. Paskin, K. J. Tauer, and R. J. Weiss, *J. Phys. Chem. Solids*, 1956, *1*, 45–60.

26. G. M. Androes, and W. D. Knight, *Phys. Rev. Letters,* 1959, *2(9)*, 386–7.
27. G. M. Androes, and W. D. Knight, *Phys. Rev.,* 1961, *121*, 779–87.
28. K. D. Nelson, Z. Q. Mao, Y. Maeno, and Y. Liu, *Science*, 2004, *306*, 1151–4.
29. A. P. Mackenzie, and Y. Maeno, *Rev. Mod. Phys.,* 2003, *75*, 657–712.
30. F. London, and H. London, *Proc. Roy. Soc. A*, 1935, *149*, 71–88.
31. See, for example, M. Tinkham, *Introduction to Superconductivity*, second edition, 1996, Dover, New York; pp 100–108. Tinkham give an estimate of 500 Å (5 μm) as a typical value of λ, p. 2 footnote 4.
32. F. London, *Phys. Rev.*, 1948, *74(5)*, 562–73.
33. A. Yu. Sadykova, A. S. Saykaeva, A. V. Kostochko, A. N. Glebov, and V. G. Moozyukov, *Int. J. Quantum Chem.*, 1992, *44*, 935–47.
34. See, e.g., the list assembled at Fermi Lab: http://www-d0.fnal.gov/hardware/cal/lvps_info/engineering/elementmagn.pdf
35. J. C. Slater, *Quantum Theory of Matter*, second edition, 1968, McGraw-Hill, New York; pp. 234–242.
36. W. S. Corak, and C. B. Satterthwaite, *Phys. Rev.,* 1956, *102*, 662–6.
37. R. Radebaugh, and P. H. Keesom, *Phys. Rev.*, 1966, *149*, 209–16.
38. A. T. Hirshfeld, H. A. Leupold, and H. A. Boorse, *Phys. Rev.*, 1962, *127*, 1501–7.
39. R. J. Trainor, M. B. Brodsky, *Phys. Rev. B*, 1975, *12*, 4867–9.
40. H. R. O'Neal, and N. E. Phillips, *Phys. Rev.*, 1965, *137*, A748–59.
41. W. H. Keesom, and J. A. Kok, *Akademie der Wetenschappen, Amsterdam, Proceedings*, 1932, *35*, 743–48.
42. C. A. Bryant, and P. H. Keesom, *Phys. Rev.*, 1961, *123*, 491–9.
43. H. R. O'Neal, and N. E. Phillips, *Phys. Rev.*, 1965, *137*, A748–59.
44. W. H. Keesom, and J. A. Kok, *Physica*, 1934, *1*, 503–12.
45. A. T. Hirshfeld, H. A. Leupold, and H. A. Boorse, *Phys. Rev.*, 1962, *127*, 1501–7.
46. A. Brown, M. W. Zemansky, and H. A. Boorse, *Phys. Rev.*, 1952, *86*, 134–5.
47. C. Buzea, and K. Robbie, *Supercond. Sci. Technol.*, 2005, *18*, R1–R8.
48. P. Strange, *Relativistic Quantum Mechanics: With Applications in Condensed Matter and Atomic Physics*, Cambridge University Press, 1998, London; pp. 317–318.
49. W. J. de Haas, P. M. van Alphen, *Proc. Netherlands Roy, Acad. Sci.*, 1930, *33*, 1106.
50. Referenced by, D. Shoenberg, *Magnetic Oscillations in Metals*, 1984, Cambridge University Press, London; p. 2.
51. W. J. de Haas, and P. M. van Alphen, *Proc. Netherlands Roy, Acad. Sci.*, 1930, *33*, 1106; Fig. 6, p. 1116.
52. J. H. Condon, *Phys. Rev.*, 1966, *145*, 526–35; from Fig. 10, p. 532.
53. For leading references see: A. Audouard, C. Jaudet, D. Vignolles, R. Liang, D. A. Bonn, W. N. Hardy, L. Taillefer, and C. Proust, *Phys. Rev. Letters*, 2009, *103*, 157003 1–4.

10 Quantum Hall Effect

Two-dimensional electron systems were well known in the 1960s prior to the discovery of the quantum Hall effect [1]. Using these systems, investigators explored a wide variety of effects, leading to publication of magnetoconductance oscillations in 1972 [2] (see Figure 10.1). Figure 10.1 is an illustration from the one-page paper by Ando, Kobayashi, Komatsubara et al., in which they reported the transverse magnetoconductance of a two-dimensional electron gas [2]. The experimental work was conducted by Kobayashi and Komatsubara on "n-type layers on Si(100) surfaces" [2].

Figure 10.2 presents the data from the same authors on the transverse conductivity versus the Landau level index. The Landau level index is a quantum index in Landau's quantum mechanical treatment of the electron gas as a harmonic oscillator [3], which will be discussed in this chapter.

Figure 10.2 shows that the authors had developed an early model for the quantum of magnetoconductance, e^2/h. Although it was not realized as such at the time, this appears to be the first literature report of a quantum of magnetoconductance/magnetoresistance. This report provides definitive experimental support for the existence of magnetoconductance quanta. Independent experimental evidence has subsequently been generated for the existence of electric conductance/resistance quanta, which differ from the magneto-counterparts by a factor of 2 or 0.5, respectively (see Chapter 6). The paper by Ando, Kobayashi, Komatsubara et al. stimulated a substantial amount of theoretical work on electron scattering in two-dimensional electron systems, some of which we will examine after a brief discussion of Landau levels.

From the point of view of superconductivity, the most important feature of the quantum Hall effect, QHE, is the discovery of the quantum of magnetoresistance:

$$\frac{h}{e^2} = 25,812.807\Omega \tag{10.1}$$

This discovery illustrated that resistance is a quantized feature of the quantum mechanics of conduction in the presence of an external magnetic field. This value is precisely twice the value of the reciprocal of the quantum of electric conductance. Quantum of electric conductance values have been determined by a number of methods, including conductance in point contacts [5], carbon nanotubes [6,7], and graphene [8] (see also Chapter 6).

In the original paper on the quantum Hall effect, the number of states within each Landau level was given by [4]

$$N_L = \frac{eB}{h} \tag{10.2}$$

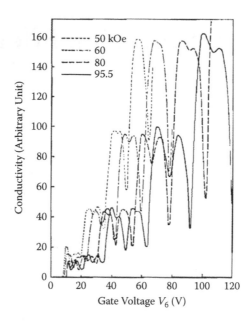

FIGURE 10.1 Magnetoconductivity oscillations, sample 1, versus gate voltage, $T = 1.4$ K. (From T. Ando, Y. Matsumoto, Y. Uemura, M. Kobayashi, and K. F. Komatsubara, *J. Phys. Soc. Japan*, 1972, *32*, 859.)

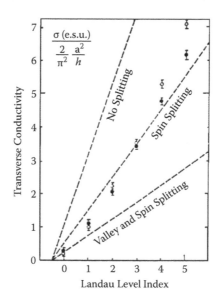

FIGURE 10.2 Peak values of transverse conductivity versus Landau level indices; same details as Figure 10.1. (From T. Ando, Y. Matsumoto, Y. Uemura, M. Kobayashi, and K. F. Komatsubara, *J. Phys. Soc. Japan*, 1972, *32*, 859.)

The Hall conductivity, which is simplified by zero scattering, is given by [9].

$$\sigma_{x,y} = -\frac{Ne}{B} \qquad (10.3)$$

When a Landau level is fully occupied and $N = N_L i$ ($i = 1,2,3,\ldots$), the Hall conductivity becomes

$$-\sigma_{x,y} = \frac{ie^2}{h} \qquad (10.4)$$

which was the first formal expression of the quantum Hall effect [4]. In Equation (10.4), e^2/h is the quantum of magnetoresistance, and i is an integer in the Landau treatment. Regardless of the formulation, the relationships in Equations (10.1) and (10.4) show more clearly than any previous argument that both magnetoconductance and magnetoresistance in electronic quantum mechanical systems are a sole function of the quantum mechanics of the system. This conclusion follows directly from the derivation of the quantum of magnetoresistance, the von Klitzing constant, Equation (10.1), which started with the Landau levels of electrons in the Hall effect experiment [4].

Landau levels were developed by Lev Landau, starting with wave functions for an electron gas that did not utilize atomic quantum numbers, let alone consider the impact of electron spin or the vibrational coordinates of atomic nuclei on the problem [3]. Landau was working with an electronic version of simplified quantum mechanics for conduction electrons in metals [10]. Since superconductivity involves loss of resistivity in metals, it is essential to be clear about the origins of resistivity if we hope to be accurate in a description of the change that brings on its disappearance. In this regard it is significant that the original paper on BCS theory [11] mentions resistance only twice and makes no mention of the events causing disappearance of resistance.*

Quantum Hall effects are also important to understanding superconductivity because quantum Hall systems provide clear examples of quantum mechanical control of electromagnetic properties of conducting systems in a somewhat less overarching construct than that offered by superconductivity. Quantum Hall effects are seen under conditions that could support superconductivity with the exception of (1) two-dimensional conductor structure and (2) external magnetic field. Integer quantum Hall effects are typically observed with two-dimensional electron conductors (conducting films, appropriately supported) in strong magnetic fields, generally at low temperatures, 4 K and below. Josephson junctions are another example of ~two-dimensional analogs of superconducting systems.

---

* [11] p. 1175, column 1, first sentence second paragraph, "When superconductivity was discovered by Onnes (1911), and for many years afterwards, it was thought to consist simply of a vanishing of all electrical resistance below the transition temperature."; p. 1197, table IV, note, reference to Pippard's method, "skin resistance."

LANDAU LEVELS

Electron cyclotron resonance in metals is the foundation for the Hall effect and the quantum Hall effect. At age 22, Lev Landau published a paper titled, "Diamagnetism of Metals," [12] in which he solved the quantum mechanical problem for charged particles in a uniform magnetic field. Diamagnetic orbital angular momentum is a quantum property of electrons with or without consideration of spin. Orbital diamagnetism arises from the electron charge. Since electrons carry a negative charge, the right-hand rule tells us that the magnetically induced orbital motion of the electron produces a magnetic field opposed to the original field, that is, a diamagnetic current. You will recall that most elemental metals have a magnetic susceptibility that is paramagnetic, which indicates that electron spin and other paramagnetic components dominate the magnetic susceptibility for most metals. For the 14 diamagnetic elemental metals in the periodic table, it is likely that Landau diamagnetism is the source of their diamagnetic response to external magnetic fields. Note that Table 9.1 shows only 13 diamagnetic metals because bismuth, Bi, the most diamagnetic element was not included in that table.

The model Landau used is free electron gas in a uniform magnetic field. His title indicates that this is his model for electron cyclotron motion in a metal. Landau's model has led to many insights. Landau levels are observable in metals because they function on a scale that is sufficiently larger than the atomic lattice constant, so that atomic scale features like the Fermi surface of atoms or bonding pockets of electron density [13] do not significantly perturb relative conducting electron energies in the system.

Landau's analysis begins with cyclotron resonance of a thermal electron gas in the presence of a uniform magnetic field.

$$\omega_c = \frac{eB}{m_e} \tag{10.5}$$

This is the defining equation for electron cyclotron resonance. ω_c is the electron cyclotron frequency. m_e is the electron mass, e is the electron charge, and B is the magnetic flux density. In Landau's original electron gas model, there is no magnetoresistance. This is because there is no magnetoresistance in a low-pressure plasma. Atomic nuclei that are wave-mechanically associated with moving electrons are essential for partial wave scattering. For simplicity, the effects of electron spin and the z-component of the orbital angular momentum are also often not mentioned. In any case, solutions for the energy levels have the form of a simple harmonic oscillator.

$$E_c = \left(n + \frac{1}{2} \right) \hbar \omega_c \tag{10.6}$$

When n is the index that labels (quantizes) the Landau levels. Complete derivations of Equation (10.6) (generally with different notation) and elaboration of the details can be obtained from a number of sources, including Yoshioka [1b], Marder [14], and Ezawa [15]. Landau levels of electrons in magnetic fields are useful in explaining a wide scope of magnetic properties of metals, not the least of which is the quantum Hall effect.

Before we leave Landau's treatment we need to examine the cause of positive Hall coefficients within the context of this framework. Positive Hall coefficients would be easily explainable if there were positively charged particles that moved with the momentum of an electron as part of a metal structure. The only experimentally documented positively charged particles that we are aware of in normal metals are atomic nuclei. As a consequence, we need to look at electrons as a possible source of positive Hall coefficients.[*]

ORIGINS OF POSITIVE HALL COEFFICIENTS

Positive Hall coefficients are important primarily in the study of p-type semiconductors. Among the common metallic elements, Mn, Zn, Cd, diamond lattice Sn,[†] and Pb have positive Hall coefficients at 300 K [16]. Classical equations for the Hall coefficient require a charge moving with the mobility of an electron in the metal. The equations do not require that the charge be positive or negative. They do require that the direction of motion corresponds to a negative Hall current density to obtain a positive Hall coefficient.[‡]

For a conductor that has a potential difference between the two poles, many people have adopted the quasiparticle model for a current that has an electron enter the metal at the negative pole and a positive hole enter the metal at the positive pole. This part of the model satisfies the continuity equation for the metal.[§] It also introduces a positive hole with the mobility of an electron into the metal, a real contribution to the system for which there is no experimental evidence. In our effort to locate experimental evidence concerning electron hole interactions in condensed matter, the only real evidence we located was in a paper titled, "Experimental Observation of Electron–Hole Re-collisions" [17]. This paper discusses intense laser field formation of electron hole pairs in a gallium arsenide quantum well and their subsequent decay through a process that includes electron recollision with the ionized nucleus in the quantum well. These hole features are explicitly referenced to the atomic core and could not be utilized to describe the origin of positive Hall coefficients.

---

[*] A positive Hall coefficient is a consequence of a negative Hall current density, $j_y < 0$ (see Equation [9.2] and the following discussion).

[†] Marder [16] gives the Hall coefficient for β-tin, the tetragonal allotrope. It is α-tin, the diamond lattice allotrope that has a positive Hall coefficient. α-Tin is metastable at 300 K.

[‡] The equation for the Hall coefficient,

$$R_H = -\frac{E_x}{Bj_y},$$

shows it to be given by the ratio of the electric field on the x-axis to the product of the magnetic flux density and the Hall current density on the y-axis. Positive Hall coefficients appear when the Hall current density is negative.

[§] If the charge associated with a positive hole is specifically referred to an atomic core in the lattice, there is no basis for using a positive particle to solve the problem of positive Hall coefficients. If the positive charge associated with a quasiparticle hole has a real + charge, then all of the core charges in the lattice, associated with the positive quasiparticles, must be reduced by one unit of positive charge per quasiparticle to maintain charge continuity. All of the electrons must be treated identically, and the number of electrons just matches the nuclear charge in the lattice for a neutral metal.

If positive charges moving with the mobility of electrons were part of the structure of tin, for example, there would be experimental evidence for the existence of those charges and particles. There is none that we can find. As a consequence we must look elsewhere for a coherent explanation of positive Hall coefficients.

Landau levels derived using all of the magnetic properties of the system hold the key to this problem. Adding the z-component of electron orbital angular momentum increases the complexity of both the Landau diamagnetism problem and the solution. A solution of this type appears in Yoshioka's excellent book [18]. Our choice of coordinates is indicated in Figure 10.3.

A full quantum mechanical Hamiltonian for the Landau levels in the 2D electron gas is given with $q = -e$.

$$H = -\frac{\hbar^2}{2m}\nabla_2^2 - \frac{q}{2m}BL_z + \frac{q^2B^2\rho^2}{8m} \tag{10.7}$$

The interaction of the magnetic moment with the external magnetic field is

$$L_z = \frac{\hbar}{i}\frac{\partial}{\partial\phi} \quad \text{and} \quad \mu_z = \frac{q}{2m}L_z.$$

$$-\vec{\mu}\cdot\vec{B} = -\mu_z B = -\frac{q}{2m}L_z B \tag{10.8}$$

where L_z commutes with the Hamiltonian, the wave function can be written as

$$\Psi = R(\rho)e^{i\nu\phi} \tag{10.9}$$

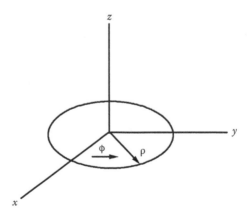

FIGURE 10.3 Coordinates used in the full description of the Landau levels for a 2D electron gas.

where $v = 0, \pm1, \pm2, \ldots$ With this wave function, the Schrödinger equation can be written as

$$\left[-\frac{\hbar^2}{2m}\nabla_2^2 + \frac{1}{2}m\omega^2\rho^2 - (E - \hbar\omega v) \right]\Psi = 0 \tag{10.10}$$

The Laplacian operator is expanded to

$$-\frac{\hbar^2}{2m}\left[\frac{1}{\rho}\frac{\partial}{\partial\rho}\left(\rho\frac{\partial}{\partial\rho}R \right) - \frac{v^2}{\rho^2} \right] + \frac{1}{2}m\omega^2\rho^2 R - (E - \hbar\omega v)R = 0 \tag{10.11}$$

Solutions for these equations are available through algebraic manipulations and the use of Mathematica. The eigen function obtained

$$\Psi_{sv}(\rho,\phi) = z^{|v|}e^{-\frac{1}{2}z^2}F\left(-s, 1+|v|, -\frac{1}{4}z^2(1+4s+2|v|) \right)e^{iv\phi} \tag{10.12}$$

where

$$z = v\frac{\rho}{\sqrt{2l}},$$

and l, the Larmor length, is,

$$l = \sqrt{\frac{\hbar}{eB}}.$$

In this solution, the energies of the Landau levels are given by

$$E = \hbar\omega_c\left(\frac{1}{2} + \frac{|v|+v}{2} + s \right) \tag{10.13}$$

where the quantum numbers v and s are restricted to the following values:

$$s = 0, 1, 2, \ldots$$

$$v = 0, \pm1, \pm2, \ldots$$

Landau energies obtained from Equation (10.13) are the same as those obtained from Equation (10.6), which means that,

$$n = \frac{|v|+v}{2} + s.$$

The Landau ground state has the quantum number $\eta = 0$ and is highly degenerate, as are the levels with $\eta > 0$. In the Landau ground state, the only allowed values for the

z-component of the electron orbital angular momentum quantum number, ν, are negative or zero (see Equation [10.13]). This appears to be the factor that creates the negative Hall current in materials with a positive Hall coefficient. In this special case the electron current is directed by electron angular momentum and not an electric potential.

For higher Landau levels, ν can have any of its potential values: positive, negative, or zero. If we assume that the populated states are dominated by wave functions with the fewest nodes, that is,. $s = 0$, then only positive ν are allowed. All that is required is dominance of positive values of ν for the populated states and the current density on the y-axis will be positive, leading to a negative Hall coefficient, R_H,

$$R_H = -\frac{E_x}{Bj_y} \qquad (10.14)$$

The coordinates for the Hall experiment can be seen in Figure 10.4. This illustrates classical cycloidal motion in a Hall experiment to produce a negative Hall coefficient. In a classical picture, the current along the y-axis is entirely driven by the Hall effect in a pure metal, current and fields following physics conventions. In this example, the y current density, j_y, is positive, so the Hall coefficient is negative.

It is well known that high electron mobility and a high charge carrier density both favor negative Hall coefficients in agreement with the expectations of the Landau level model. It is interesting that the element with the largest positive Hall coefficient at 300 K is manganese, Mn [19]. Manganese also has the highest resistivity of any elemental metal (see Table 8.1, which is relevant).

ORIGINS OF ELECTRON SCATTERING IN METALS

Since we are introducing the notion of magnetoresistance in metals as the manifestation of dissipative electron scattering caused by magnetic fields, it would be a good idea to explain why we think this is true. Dissipative electron scattering in metals is known to be caused by application of an external electric field. Classically speaking, an applied electric field is an accelerating field for conduction band electrons in the metal. Dissipative electron scattering in metals can also be caused by application of an external magnetic field. This dissipative electron scattering is, in fact, magnetoresistance. Classically speaking, the applied magnetic field causes a centripetal acceleration of any electron with motion that crosses the field. The clearest example of pure magnetoresistance that we are aware of is the electron effective mass that can be observed in cyclotron resonance experiments at ambient temperatures.

Landau's equations for the energy levels of an electron gas are fully functional for the description of conditions in the quantum Hall regime. Landau's electron gas equations are founded on the equation for electron cyclotron resonance of an electron gas (Equation [10.5]). The medium for the quantum Hall regime is generally a cold film of metal. The fact that Equation (10.5), the cyclotron equation, is fully functional in this medium indicates that there is no need to invoke the electron effective mass, which is normally used when the cyclotron medium is a

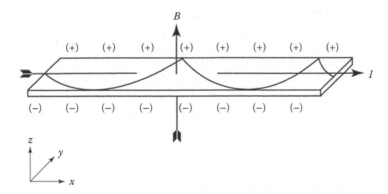

FIGURE 10.4 Hall effect in a pure metal, current and fields follow physics conventions. (*Note:* As it happens, the physics and chemistry conventions for the directions of currents and physical fields give the same results in this special case.) In this example, the y current density, j_y, is positive, so the Hall coefficient is negative.

metal at ambient temperature.[*] The one qualitative difference that connects the two conditions is the lack of dissipative electron scattering in the quantum Hall regime. In the quantum Hall regime, dissipative electron scattering is turned off by lowering the temperature. Nothing else has been shown to be a requirement for entering the quantum Hall regime, if you have an appropriately thin sample of material that is capable of nondissipative electrical conduction.[†] Since the electron mass in the quantum Hall regime is the standard value of the electron mass, it follows that dissipative electron scattering caused by the presence of an external magnetic field in a cyclotron resonance experiment is the probable agent responsible for the observation of electron effective mass.

Zero electron scattering is a qualitative difference between the full quantum Hall regime and a normal metal that has the potential to account for the observed changes in electron effective mass. Dissipative electron scattering, caused by magnetoresistance, under ordinary electron cyclotron resonance conditions will reduce the observer's value for k, the electron wave vector, for an electron observed in a cyclotron resonance experiment. Effects of the reduction are referred to as electron effective mass.

A change from electron effective mass to the real mass of the electron on entering the quantum Hall regime is inextricably associated with the end of electron scattering. We can be quite clear that the end of the bulk of electron scattering has its origins in electron quantum mechanics and not the properties of the lattice such as phonons. We know of no mechanism for abruptly changing scattering of phonons and electrons as a function of temperature per se. Such a mechanism would be an absolute essential if electron–phonon scattering were the source of the resistivity that vanishes in the quantum Hall regime and/or superconductivity as a strict function of temperature alone.

---

[*] See the section entitled Dissipative Magneto Resistance in Chapter 6 and Equation (6.5).
[†] This requirement is equivalent to saying that the material would be a superconductor if it met the other physical requirements for superconductivity: sample shape, volume, temperature, etc.

Since lowering the temperature of the sample is causative for crossing the transition from normal metal to quantum Hall regime, we can infer that a very similar process is operative in the superconductivity phase transition. Dissipative electron scattering, like that observed by Kamerlingh Onnes in mercury, is driven by electric fields, and it shuts down abruptly at T_c. Abrupt changes of electron scattering of this kind are not known to involve phonons. There are no known scattering processes involving phonons and electrons that end abruptly as a function of a temperature that systematically varies in the periodic table of the elements.

LOW AND ZERO ELECTRON SCATTERING IN THE QUANTUM HALL REGIME

In 1974, Ando and Uemura provided a quantitative description of electron scattering in quantum electron transport in a two-dimensional electron system [9,20,21]. These three papers were published as a series in the order, respectively referenced 20, 21, and 9. In the second paper in this series [20], "Theory of Quantum Transport in a Two-Dimensional Electron System under Magnetic Fields. I. Characteristics of Level Broadening and Transport Under Strong Fields," the abstract reads in part:

> To see the dependence on the range explicitly, numerical calculation has been performed for the system with scatterers with the Gaussian potential. Especially in case of short-ranged ones the peak value of the transverse conductivity has been shown to be
>
> $$\left(N + \frac{1}{2} \right) \frac{e^2}{\pi^2 \hbar},$$
>
> which depends only on the natural constants and the Landau level index $\{N\}$. It has been argued from general point of view that this fact is approximately true without reference to kinds of approximations." [20] (*sic*)

In the quotation above, N represents the Landau index, which we previously labeled as n (Equation [10.6]) and (Equation [10.13]).

Computed data for an electronic scattering model at two values of the parameter, α, presented by Ando and Uemura, are illustrated in Figures 10.5 and 10.6. These figures correspond to Figures 10 and 11 in the second paper [20] in their series on the subject of electron scattering in 2D electron systems.

Ando and Uemura's model did not focus on the origins of the scattering in MOS[*] two-dimensional electron systems. They primarily spoke of short-range scatters, which would include partial wave scattering from atomic centers in the lattice as a major source of scattering in the system.

---

[*] The acronym MOS can refer to either metal oxide silicon or metal oxide semiconductor. The two meanings are often interchangeable.

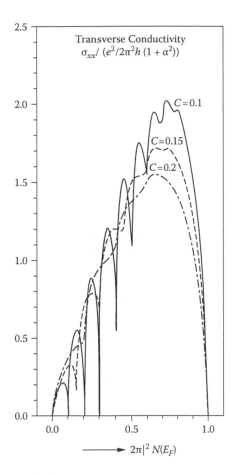

FIGURE 10.5 Transverse conductivity, σ_{xx}, versus total electrons at low concentration, $\alpha = 1.0$. (From T. Ando, and Y. Uemura, *J. Phys. Soc. Japan*, 1974, *36*, 959–67.)

It is not essential that electron scattering be exactly zero for effective observation of the quantum Hall effect, as evidenced by the observation of a quantum Hall effect in graphene at room temperature [22]. In graphene, the conducting electrons move in a combination of carbon $2p$ wave functions. Partial wave scattering from these carbon π orbitals is roughly three orders of magnitude smaller than it would be from carbon $2s$-based wave functions at the same temperature. At 300 K the partial wave scattering for unpaired electrons in carbon $2p$-basis function orbitals is not zero. In contrast to superconductivity, which requires zero electron scattering, the quantum Hall effect only requires minimal electron scattering like that found in graphene at room temperature.

In cases where electron scattering in Hall effect measurements is precisely zero, there is a substantial simplification in the equations relating to electron transport. Cyclotron behavior of the electrons in the Landau levels is stable because there is no electron scattering either from the cyclotron motion or from the current used to drive the Hall experiment. In the first paper on the quantum Hall effect, lack of

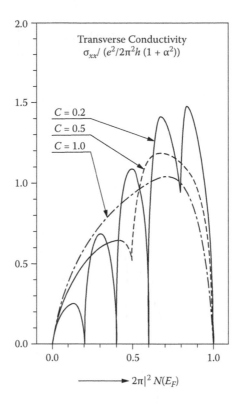

FIGURE 10.6 Transverse conductivity, σ_{xx}, versus total electrons at low concentration, $\alpha = 0.5$. (From T. Ando, and Y. Uemura, *J. Phys. Soc. Japan*, 1974, *36*, 959–67.)

electron scattering is noticed by the statement that, under conditions of no electron scattering, "the center of the cyclotron orbit drifts in the direction perpendicular to the electric and magnetic field" [23]. The quantum Hall regime shows that zero electron scattering is directly attainable in a metal system with ordinary unpaired electron spins.

QUANTUM HALL OBSERVATION

The original quantum Hall illustration from the 1980 paper by von Klitzing, Dorda, and Pepper [4] is presented as Figure 10.7. The Hall probe in Figure 10.7 was constructed from a metal oxide semiconductor field-effect transistor. Integer QHE transitions in this system are marked on the abscissa in Figure 10.7. Fractional quantum Hall transitions also appear in this figure. Fractional Hall effects were not explained until a later date and will not be covered here. They are well treated in modern texts [24].

Voltage plateaus shown in Figure 10.7, correspond to plateaus of Hall resistance that occur in units of the von Klitzing constant, h/e^2. Formally, the quantum Hall transitions can be treated as transitions between Landau levels in the system. These quantum transitions have very practical applications. One of the anticipated uses of the quantum Hall effect is the quantum resistance standard. A device that would

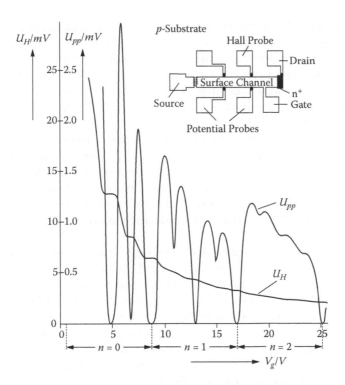

FIGURE 10.7   Hall voltage, U_H, and voltage drop between potential probes, U_{pp}, as a function of gate voltage, V_g. $T = 1.5$ K. $B = 18$ T. Source drain current, $I = 1$ μA. Device: length, $L = 400$ μm; width, $W = 50$ μm; interpotential probe distance, $L_{pp} = 130$ μm. (From K. von Klitzing et al., *Phys. Rev. Letters*, 1980, *45*, 494–7.)

produce a standard resistance related to the von Klitzing constant would be very useful in a number of applications. A device based on epitaxial graphene has received recent attention; however, we are not aware of any commercially available products at this time.

The quantum of magnetoresistance, h/e^2, and the quantum of electric conductance, $2e^2/h$, leave no doubt that both resistance and conductance are electronic phenomena without explicit involvement of the lattice of atomic nuclei. This is consistent with the Sommerfeld relationship, Chapter 6, which leads to the same inference.

PHASE TRANSITION QUANTA IN THE QUANTUM HALL REGIME

Quantum Hall regime Mott-transition initiating electric resistance quanta, $h/4e^2$, were reported in 2011 [25]. Bollinger, Božović et al. reported that their two-dimensional, epitaxially prepared, cuprate devices were switched from conduction with no resistance to an insulator state by an electric field. Electron motion under these conditions is chiral, as it is the in the experiment of Novoselov, Geim et al. [26]. Devices used by Bollinger, Božović et al., were shown to have zero resistance. They were reported to be approximately two unit cells thick [25]. Zero magnetic permeability, or −1 magnetic susceptibility would not have been obtained with these ultra-thin devices. Values for experimental magnetic susceptibilities of the devices did not appear to be available. We are confident that, if the wafers are examined with soft x-ray, magnetic circular dichroism under zero resistance conditions, the wafers will be normal metallic conductors with unpaired electron spins. Observation of zero resistance for these cuprate devices is an important discovery for our understanding of the theoretical basis for superconductivity [25].

Electric resistance quanta with a divisor of 4 first appeared in the literature in 2006 [26]. Corresponding conductance quanta, $4e^2/h$, appeared in the same paper. The report by Novoselov, Geim et al. concerns the properties of bilayer graphene in the quantum Hall regime and did not involve a phase transition. Bilayer graphene is known to carry electric currents without resistance in the quantum Hall regime. Bilayer graphene will not support superconductivity because it is not a three-dimensional conductor.

Bilayer graphene, as studied by Novoselov, Geim et al., is subject to the formation of biexcitonic Bose–Einstein correlates [27,28]. These electrically neutral charge transfer structures, in which symmetric radical anion and radical cation pairs are formed in the two layers, polarize the surfaces (see discussion of Figure 11.8). The cuprate devices that were studied by Bollinger, Božović et al. were also polarized on the Hall z-axis for the purpose of accomplishing the electric field-induced Mott transition they reported. Electrical polarization on the z-axis was explicit in the Mott transition experiment and also present in the bilayer graphene experiment [29,30]. This electrical polarization seems probable as a cause of the additional factor of 2 in the observed resistance quanta.

Quantum Hall effects are an essential component for the development of a modern analytical view of the superconducting state. Type I superconductors, which are largely restricted to elemental metals and simple alloys, are the subject of the next chapter.

REFERENCES

1. (a) K. von Klitzing, G. Dorda, and M. Pepper, *Phys. Rev. Letters*, 1980, *45*, 494–7; (b) D. Yoshioka, *The Quantum Hall Effect*, 2010, Springer, Heidelberg; Section 1.1, p. 1.
2. T. Ando, Y. Matsumoto, Y. Uemura, M. Kobayashi, and K. F. Komatsubara, *J. Phys. Soc. Japan*, 1972, *32*, 859.
3. L. D. Landau, *Z. Phys. A*, 1930, *64*, 629–37.
4. K. von Klitzing, G. Dorda, and M. Pepper, *Phys. Rev. Letters*, 1980, *45*, 494–7.
5. B. J. van Wees, H. van Houten, C. W. Beenakker, J. G. Williamson, L. P. Kouwenhoven, D. van der Marel, and C. T. Foxon, *Phys. Rev. Letters*, 1988, *60*, 848–50.
6. M. F. Lin, and K. W.-K. Shung, *Phys. Rev. B*, 1995, *51*, 7592–7.
7. L. Chico, L. X. Benedict, S. G. Louie, and M. L. Cohen, *Phys. Rev. B*, 1996, *54*, 2600–6.
8. W. Tian, and S. Datta, *Phys. Rev. B*, 1994, *49*, 5097–100.
9. T. Ando, *J. Phys. Soc. Japan*, 1974, *37*, 622–30.
10. See, e.g., M. Marder, *Condensed Matter Physics*, second edition, 2010, John Wiley & Sons, New York; Section 25.3, Free Electron Gas, pp. 769–70, and Section 25.3.2, Landau Diamagnetism, pp. 771–4.
11. J. Bardeen, L. N. Cooper, and J. R. Schrieffer, *Phys. Rev.*, 1957, *108*, 1175–204.
12. See Ref. 3. For an English translation, see, D. ter Haar, Ed., *Collected Papers of L. D. Landau*, 1965, Gordon and Beach, London; pp.31–8.

13. P. N. H. Nakashima, A. E. Smith, J. Etheridge, and B. C. Muddle, *Science*, 2011, *331*, 1583–6.

14. M. Marder, Condensed Matter Physics, second edition, 2010, John Wiley & Sons, New York; Section 25.3.2, Landau Diamagnetism, pp. 771–4.

15. Z. F. Ezawa, *Quantum Hall Effects*, second edition, 2008, World Scientific Publishing, Hackensack; Chapter 10, Landau Quantization, pp. 177–203.

16. See, e.g., M. Marder, *Condensed Matter Physics*, second edition, 2010, John Wiley & Sons, New York; Table 17.1, p. 499.

17. B. Zaks, R. B. Liu, and M. S. Sherwin, *Nature*, 2012, *483*, 510–3.

18. See ref. 1 b; 2.2.4 Landau Levels, pp. 22–25. In Yoshioka's treatment the variables have different names than we have chosen and a different set of action variables is used in solving the problem. That is, Yoshioka did not choose to quantize the z-component of the electron orbital angular momentum directly. His are, nonetheless, equivalent to ours.

19. M. Marder, *Condensed Matter Physics*, second edition, 2010, John Wiley & Sons, New York; Table 17.1, p. 499.

20. T. Ando, and Y. Uemura, *J. Phys. Soc. Japan*, 1974, *36*, 959–67.

21. T. Ando, *J. Phys. Soc. Japan*, 1974, *36*, 1521–9.

22. K. S. Novoselov, Z. Jiang, Y. Zhang, S. V. Morozov, H. L. Stormer, U. Zeitler, J. C. Maan, G. S. Boebinger, P. Kim, and A. K. Geim, *Science*, 2007, *315*, 1379.

23. K. von Klitzing, G. Dorda, and M. Pepper, *Phys. Rev. Letters*, 1980, *45*, 494–7; p. 494, column 1.

24. See, e.g., M. Marder, *Condensed Matter Physics*, second edition, 2010, John Wiley & Sons, New York; Section 25.5.2, pp. 785–791.

25. A. T. Bollinger, G. Dubuis,, J. Yoon, D. Pavuna, J. Misewich, and I. Božović, *Nature*, 2011, *472*, 458–460.

26. K. S. Novoselov, E. Mccan, S. V. Morozov, V. I. Fal'ko, M. I. Katsnelson, U. Zeitler, D. Jiang, F. Schedin, and A. K. Geim, *Nature Phys.*, 2006, *2*, 177–80; see Figure 2, p. 178.

27. J. P. Eisenstein, and A. H. MacDonald, *Nature*, 2004, *432*, 691–4.

28. S. A. Moskalenko, and D. W. Snoke, *Bose-Einstein Condensation of Excitons and Biexcitons*, 2000, Cambridge U. Press, New York.

29. E.V. Kurganova, A. J. M. Giesbers , R.V. Gorbachev, A.K. Geim, K.S. Novoselov , J.C. Maan, and U. Zeitler, *Solid State Commun.*, 2010, *150*, 2209–11.

30. A. S. Mayorov, D. C. Elias, M. Mucha-Kruczynski, R. V. Gorbachev, T. Tudorovskiy, A. Zhukov, S. V. Morozov, M. I. Katsnelson, V. I. Fal'ko, A. K. Geim, and K. S. Novoselov, *Science*, 2011, *333*, 860–3.

11 Type I Superconductivity

Our discussion of type I superconductors will be include experimental observations and theoretical models that can account for the observations. Experimental observations are the foundation of the subject and the testing ground for all theoretical descriptions.

EXPERIMENTAL TYPE I SUPERCONDUCTIVITY

Kamerlingh Onnes' 1911 discovery of the superconductivity of mercury at a temperature just below the boiling point of liquid helium [1,2] was the initiating event in this huge ongoing research effort. Data on the general subject of type I superconductivity is abundantly available, although the availability of specific details can be dependent on the trends that focus enquiry in science.

CRITICAL MAGNETIC FIELD VERSUS T FOR ELEMENTAL SUPERCONDUCTORS

Critical magnetic field, H_c, as a function of temperature, T, was carefully examined in the early literature in the effort to obtain information on the mechanism of superconductivity. A superconductor's critical magnetic field is the maximum magnetic field that it can generate as a superconductor at a given temperature. It is also the maximum external magnetic field that the superconductor can be exposed to and still maintain superconductivity. Using data from the literature, critical magnetic field as a function of temperature is plotted for a number of elemental superconductors in Figures 11.1–11.4. The plots in Figures 11.1–11.4 follow the form of those presented by Shoenberg in 1960 [3]. Data for the plots came from the references for the figures, which are collected at before the four figures.

As an aid to sorting out the periodic behavior of the critical magnetic field curves for this series of superconductors, we have reproduced the table on the T_c of one-bar elemental superconductors as Table 11.1.

When you scan Figures 11.1 through 11.4, you will notice the presence of families of curves in the data. These families of curves reflect the element-specific temperature dependence of the critical magnetic field. Existence of families of curves points to similar dependence for specific groups of elements. You can see from Figures 11.1 to 11.4 and Table 11.1 that the families of curves are explicitly associated with the periodic table. The largest family of curves includes the elements that have the electronic structure of the main group metals, the superconducting elements in columns 12 through 14 of Table 11.1 (see Figures 11.1 and 11.3). Formally, the elements in column 12, the zinc family, are transition metals; however, they have a closed d subshell, and their lowest energy conduction band is formed from singly occupied s and p atomic orbitals at the valence level. These features make the elements in column 12 electronically similar to the elements in columns 13 and 14 that also have a

TABLE 11.1

Truncated Periodic Table Showing the Bulk Elements That Are Superconducting at One Atmosphere With Their Critical Temperatures, T_c

Symbol ⟶ **Nb**
Atomic Number ⟶ 41 9.25 ⟵ Critical Temperature (T_C) in Kelvin (K)

☐ $T_C \geq 1.5$ K
▨ $1.5 > T_C > 0.1$ K
▦ $T_C \leq 0.1$ K

Group

Period	1	2	3	4	5	6	7	8	9	10	11	12	13	14	15	16	17	18
1	H 1																	He 2
2	Li 3	Be 4 0.026											B 5	C 6	N 7	O 8	F 9	Ne 10
3	Na 11	Mg 12											Al 13 1.18	Si 14	P 15	S 16	Cl 17	Ar 18
4	K 19	Ca 20	Sc 21	Ti 22 0.5	V 23 5.4	Cr 24	Mn 25	Fe 26	Co 27	Ni 28	Cu 29	Zn 30 0.85	Ga 31 1.08	Ge 32	As 33	Se 34	Br 35	Kr 36
5	Rb 37	Sr 38	Y 39	Zr 40 0.6	Nb 41 9.25	Mo 42 0.92	Tc 43 8.2	Ru 44 0.5	Rh 45	Pd 46	Ag 47	Cd 48 0.57	In 49 3.4	Sn 50 3.7	Sb 51	Te 52	I 53	Xe 54
6	Cs 55	Ba 56	La 57 6.0	Hf 72 0.38	Ta 73 4.4	W 74 0.01	Re 75 1.7	Os 76 0.7	Ir 77 0.1	Pt 78	Au 79	Hg 80 4.15	Tl 81 2.4	Pb 82 7.2	Bi 83	Po 84	At 85	Rn 86

Note: Critical temperatures, T_c in kelvin, are shown at the bottom right for each element. Period 7 (all radioactive and dominantly synthetic) and the *f* elements in period 6, with the exception of lanthanum, La, are not shown.

closed *d* subshell and form conduction bands from partially filled *s* and *p* orbitals at the valence level.

Normal metal–superconductor phase transitions are classically described as second order. This means that any point along the transition curve is formally a critical point and should be associated with a critical exponent [4]. The families of curves for critical magnetic field versus temperature will form a cluster of related critical exponents in any statistical thermodynamic treatment of superconductivity.

The second largest family of curves comes from the vanadium, V, family, Figure 11.4, with the addition of titanium, Ti, and zirconium, Zr, from Figure 11.2. Osmium, Os, and ruthenium, Ru (Figure 11.2) represent the smallest family of curves in this collection.

In the main group metal family, gallium, Ga, appears as an outlier. The critical field at ~0 K for gallium is substantially low, compared to the critical temperature, considered with other members of the main group family of superconducting elements. Gallium has a relatively weak lattice structure. Its one bar melting point of 29°C makes gallium a room temperature liquid metal in Florida in the summer. The only metals with melting points inferior to gallium are: cesium, 28.5; francium, 27; and mercury, –38.9°C. Since mercury, Hg, looks like a normal member of the main group metal family, it seems unlikely that gallium's weak lattice is the cause of its relatively low H_c. Ultimately, the cause of the low H_c for gallium must reside in the density of superconducting states at the lowest temperatures. This density of states will translate to an altered critical exponent for the critical magnetic field versus temperature curve for gallium as a superconductor.

Niobium, Nb, has both the highest superconducting critical temperature and critical magnetic field at ~0 K of the elemental superconductors. It should not be

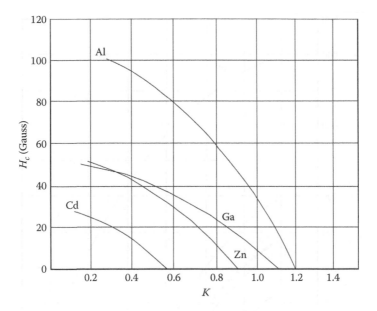

FIGURE 11.1   Critical magnetic field, H_c, v. T, low T_c, main group metals.

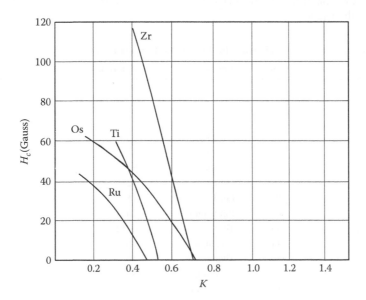

FIGURE 11.2   Critical magnetic field, H_c, v. T, low T_c, vanadium–tantalum family and iron family.

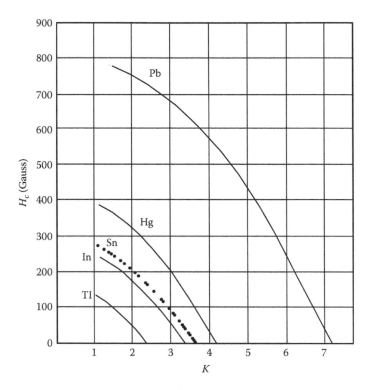

FIGURE 11.3 Critical magnetic field, H_c v. T, high T_c, main group metal family. Data points included for tin, Sn (from R.W. Shaw et al. *Phys. Rev.*, 1960, *120*, 88–91) to illustrate the error features of these plots.

References for Figures 11.1 through 11.4

Symbol	Name	Reference
Al	Aluminum	7
Cd	Cadmium	7
Ga	Gallium	7
Hg	Mercury	8
In	Indium	8
Nb	Niobium	9
Os	Osmium	10
Pb	Lead	11
Ru	Ruthenium	10
Sn	Tin	6, 12
Ta	Tantalum	13
Ti	Titanium	14
Tl	Thallium	8
V	Vanadium	15
Zn	Zinc	7
Zr	Zirconium	16

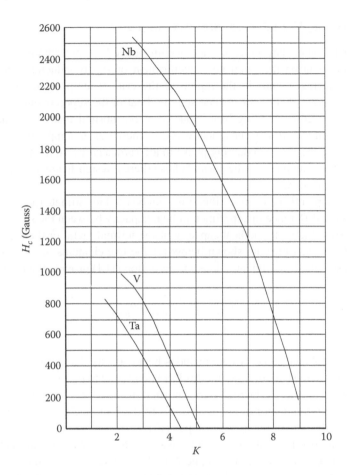

FIGURE 11.4   Critical magnetic field, H_c, v. T, high T_c, vanadium–tantalum family.

surprising that niobium is a Type II superconductor [9,17] and has both a lower and upper critical magnetic field, H_{c1} and H_{c2} respectively.

In dealing with the data on elemental superconductor critical magnetic field as a function of temperature, it is tempting to utilize a reduced scaling to present the data. Scaling can be reduced by dividing the experimental values for temperature by the critical temperature at the superconductivity phase transition, T_c, or it can be further reduced by dividing temperature by T_c^2. Similarly the scale for critical magnetic field can be reduced by dividing experimental values by the critical field at ~0 K, $H_{c,0}$. The downside of this approach to data presentation is that it reduces our ability to visually estimate errors. It may also reduce the possibility for developing a theoretical explanation for either T_c or H_c, because the atomic structure-based differences in these parameters are hidden from view by the reduced scaling. If you use reduced scaling on both axes, the curves in Figures 11.1 to 11.4 will all appear to be quite similar. In reality, data for a number of elements is of sufficient quality to be quite clear that the critical magnetic field curves for these elements versus T have distinct shapes that a theoretical model should account for in detail.

THERMODYNAMIC VARIABLES AND SUPERCONDUCTIVITY

In our discussion of magnetoresistance, we introduced a representative selection of the available data on the electronic specific heat of superconductors and the corresponding normal metals in the region of the critical temperature, T_c. The data consistently shows that the electronic specific heat in the superconducting phase is larger than the corresponding heat capacity in the normal phase at a marginally higher temperature. In materials, heat capacity increases as temperature increases. This is true for the superconducting phase in all known cases. In the transition from the superconducting phase to the normal phase there is a significant loss of electronic specific heat capacity.

Data in Chapter 9 and from the literature can be used to estimate the magnitude of the ratio of the difference in experimental heat capacity between the superconductor and the normal metal, to the heat capacity of the normal metal or the superconductor all at T_c. These ratios are shown below for the five elements for which we have located these data. Experimental data and Equation (11.1) provide the results shown in Table 11.2

$$\frac{\Delta C_{P,\,sc-n}}{C_{P,\,n}} = \frac{C_{P,\,sc} - C_{p,\,n}}{C_{p,\,sc}} \tag{11.1}$$

Experimental heat capacity data includes heat capacity due to the magnetic state of the system. In the case of niobium this has been found to be a significant contributor to the heat capacity of the high purity normal metal at very low temperature [17]. Magnetic perturbations for niobium arose from the high nuclear magnetic moment associated with some of the metal isotopes [17].

Change in specific heat capacity in the superconductor to normal-metal phase transition reflects the changes in conduction–electron scattering and changes in the system grand wave function that happen during the same transition. In the superconductor, all of the conducting electrons are thermal energy carriers. Increasing heat capacity in the transition from the normal metal to the superconductor suggests that the scattered electrons, or some fraction of the scattered electrons, in the normal state are not effective carriers of thermal energy in the system. Because of this, there

TABLE 11.2
ΔC_P Ratios at T_c for Elemental Superconductors

Symbol	Name	$\Delta C_P/C_{P,n}$	T_c (K)	$\Delta C_P/C_{P,sc}$	References
V	Vanadium	0.94	5.4	0.48	18
Nb	Niobium	0.70	9.25	0.41	19
Tc	Technetium	0.85	8.2	0.46	20
Sn	Tin	0.42	3.7	0.30	21,22,23
Tl	Thallium	0.13	2.4	0.11	24

Note: ΔC_P refers to $C_{P,sc} - C_{P,n}$ at T_c.

is an increase in the number of thermal energy carriers in the superconductor as compared to the corresponding normal metal at T_c.

Changes in specific heat capacity at T_c are connected to changes in magnetic susceptibility of the phase by their mutual dependence on the system grand wave function. Magnetic susceptibility of the normal conductor phase is in the general range of, $\chi \approx \pm 10^{-5}$. The sign of the susceptibility is positive for paramagnetic metals and negative for diamagnetic metals. For the bulk superconducting phase, the magnetic susceptibility is -1, a perfect diamagnet. This state is attained with the end of electron scattering resistivity at T_c. Electron scattering is a mechanism for communicating local electron spin response to an external magnetic perturbation to distant parts of the Condon domain. When this ceases at the critical temperature the possibility of communicating local spin response to external magnetic perturbations to other parts of the domain goes to zero. Loss of definition of the magnetic domain for the normal metal is compensated for by definition of the magnetic susceptibility of the bulk superconductor sample through the existence of a critical magnetic field and the Meissner Ochsenfeld effect on the superconductor side of the phase transition.

As far as we have been able to determine there is no experimental evidence to support the idea that electron spin pairing is the foundation of the perfect diamagnetic domain. There is evidence that the complete absence of electron scattering in the conduction band is directly associated with the superconducting state of perfect diamagnetism. If electron spin pairing was the foundation of a perfect diamagnetic domain, the magnetic susceptibility of an ultra-pure hydrocarbon like normal hexatriacontane, $n\text{-}C_{36}H_{72}$, a straight-chain, saturated, waxy hydrocarbon, would be closer to -1. As it is, all of the known closed-shell, saturated C, H and/or O molecules that we have found magnetic susceptibility data for are diamagnetic with volume magnetic susceptibilities in the approximate range of $\sim 10^{-5}$. Zero magnetic permeability of superconductor the Meissner Ochsenfeld effect, is known to be the source of perfect diamagnetic susceptibility, $\chi = -1$.

Magnetic susceptibility for spin free, easily purified, closed-shell molecules, such as methyl-cellulose [25], $\chi = -102 \bullet 10^{-6}$ cm^3/mol, shows that there is sufficient spin response to an external magnetic field, and communication of that response within the molecule and/or domain, that the magnetic susceptibility of methyl-cellulose is in the normal range for weak diamagnets. It is the lack of electron scattering, with the associated exchange of spin response information, that creates perfect diamagnetism at the transition temperature to superconductivity. Magnetic susceptibility in metals is generated by domain-wide communication of local electron spin response to perturbations from an external magnetic field. Zero communication of the response throughout the domain will have the same effect as if there was zero response in the first place. Zero response will also end the definition of the domain, which is taken up for the bulk superconductor by the critical magnetic field.

Starting from the observation that the end of electron scattering must be involved in the superconducting phase transition *per se*, we adapt and extend London's suggestion [26] that the exchange interaction, mediated by electron scattering, is central to the communication of electron response within a normal metal magnetic domain. In the absence of electron scattering, this type of exchange interaction goes to zero and there is no communication of electron spin responses to other parts of the domain.

Perfect diamagnetism for the superconductor is insured on the superconductor side of the phase transition by the existence of a critical magnetic field associated with the Meissner–Ochsenfeld effect (see the section on "Thermodynamic Requirements for a Superconductor Energy Gap").

THE MOTT TRANSITION, ELECTRON PAIR BONDS, AND SUPERCONDUCTIVITY

Sir Nevill Mott's widely quoted 1949 paper [27] introducing what is now known as the Mott transition, concerns a phase transition caused by electron spin pairing that converts a metal to an insulator or the reverse. Mott's understanding of the bonding in both metals and insulators was profound.

Mott was an unusual scientist, in that he had a deep understanding of chemical bonding as known by chemists [28]. It is historically and scientifically significant that Linus Pauling dedicated all of the editions of his book on the chemical bond to G. N. Lewis. Lewis is seen by many, as the scientist who introduced the concept of the electron-pair chemical bond [29]. Two spin-paired electrons are capable of maintaining an increased electron density between two atomic nuclei. This increase in electron density between the nuclei has the effect of holding the nuclei together. It also has the effect of reducing the mobility of the spin-paired electrons as compared to the mobility of two electrons with unpaired spins in individual orbitals. Ability of spin-paired electrons to bond atomic nuclei is a consequence of what may be thought of as reduced interelectron repulsion resulting from spin pairing.[*] Mott received the Nobel Prize in 1977.

Quantum mechanical foundations for understanding electron-pair chemical bonding are subtle and deep. These foundations begin with the structure of the hydrogen molecule, H_2, which in many ways is the archetype for all of the nonionic electron-pair bonds. Distinguishing features of electrons in electron-pair bonds and single electron bonds in metal conduction bands are: (1) in electron pair bonds the magnetic coupling is between the two bonding electrons; and (2) the magnetic coupling that supports metallic bonds occurs between the bonding electron in the conduction band and inner shell unpaired electrons on bonded atoms. The presence of either of these two types of magnetic coupling of bonding electrons is essential for generating strongly bonded molecular systems. Current generation LCAO SCF-MO[†] programs

---

[*] Spin pairing permits two electrons to occupy the same wave function. In a so-called bonding orbital, electron density between adjacent nuclei that share the orbital increases because of both constructive interference between the basis wave functions and the effective decrease in electron repulsion associated with spin pairing. The negative enthalpy associated with spin pairing is not a decrease in electron repulsion but it decreases the positive enthalpy contributions of factors like interelectron repulsion that destabilize chemicaal bonds. Coulombic nuclear attraction associated with the increase in electron density between atomic centers is the cause of chemical bond formation.

[†] Refers to linear combination of atomic orbitals self-consistent-field molecular orbital, LCAO SCF-MO, theory.

are sufficiently robust to demonstrate this, at least for the electron pair bond, using any legitimate test that one would care to construct [30,31].*

The problem in delocalization of paired electrons arises from the fact that extensive delocalization of electron pairs is only known in what are described as "alternant hydrocarbon π-electron systems." This piece of jargon comes from molecular orbital theory in organic chemistry. Alternant hydrocarbons are hydrocarbons in which there are alternating single and double electron-pair bonds. "Double bonds," which are common in organic chemistry, cannot exist in the σ-bonded electronic framework of a metal like tin. Bonding in tin will not support π-bonds, which are essential for forming double bonds. The reason for this bonding failure stems from the relative bond enthalpies of the σ and π bonds in the tin metal system. In tin, π bonds are unknown. Pi (π) bonding is only known for elements in the second and third rows of the periodic table. Formation of strong homonuclear π-bonds is more or less restricted to just carbon and nitrogen. Multiple metal–metal bonds are known [32]; however, all of the known examples require the support of an intricate organic chemical ligand structure for stability. There are no systems of alternating single double bond structures in metallic conduction, other than possibly magnesium diboride, MgB_2,[†] that we are aware of.

Sigma (σ) bonded conductors, like tin metal, depend upon a qualitatively different form of chemical bonding from the electron-pair, alternate hydrocarbon model. Bonding in metallic conductors was introduced and discussed in Chapter 7. It can be understood as second-order bond formation in a perturbation molecular orbital model. Second-order bond formation explicitly avoids formation of an energy gap in formation of a bond that expands the size of a metal cluster. Formation of an energy gap is a characteristic of bonding in both insulators and superconductors. There is a distinct origin for the gap in these two cases. In insulators, the gap arises from first-order chemical bond formation. In superconductors the gap arises from: elimination of electron scattering resistance to electrical conduction, so the chemical potential of the superconducting electrons drops by the free energy corresponding to. This first component of this model of the energy gap is an effect of the change of entropy associated with the loss of dissipative electron scattering in the superconducting phase transition. the change in electron entropy between the normal conductor which has dissipative electron scattering and the superconductor which does not.

---

* Modern MO programs for dealing with one electron metallic bonds have not yet bloomed as fully as their electron pair MO counterparts. It is reasonable to anticipate the development of new strategies for dealing with the astronomical number of potential magnetic interactions in metals so that the two bonding systems will once again be on a theoretical par.

† Bonding in the MgB_2 structure must be open shell, and contain unpaired electron spins. If the bonding in the boron layers were even-electron, spin-paired, like it is in graphene, MgB_2 would not be a three-dimensional conductor by analogy to graphite.

EXPERIMENTAL APPROACHES TO SPIN PAIRING AND SUPERCONDUCTIVITY

Physical evidence that superconductors must be conductors comes from the allotropes of tin. White tin, the β allotrope of the metal is a conductor at ambient temperatures and a superconductor at temperatures below 3.72 K. White tin has a tetragonal lattice, distorted cubic closest packing lattice structure.[*] Grey tin, the α-tin allotrope, is thermodynamically stable at temperatures below 13.2° C. It is a semiconductor at ambient temperature and is not known to be a superconductor under any conditions. Most of the bonding in grey tin is electron spin paired. The crystal structure of grey tin is the diamond lattice, as compared to the distorted face-centered cubic closest packing structure of white tin. In the physics literature, the crystal structure of grey tin is on occasion referred to as tetragonal. This designation chooses a different set of axes for the unit cell than the diamond lattice designation. The diamond lattice unit cell has the virtue of identity with the crystal structures of other members of column 14[†] of the periodic table.

Empirical validation of spin pairing in superconductors might involve use of innovative experiments. In principle, it would be possible to determine the electron spin state of a superconductor in the quantum Hall regime. This could be accomplished using soft x-ray magnetic circular dichroism and Hall bars similar to those studied by Bollinger, Božović et al. [34]. Early studies of the Knight shift in superconducting tin were almost certainly conducted in the quantum Hall regime (see Chapter 4). The data collected under those conditions showed the conductor to be capable of producing a Knight shift below the superconducting critical temperature that was different from that seen above the same temperature. This data is most easily explained if tin is an unpaired spin conductor in the quantum Hall regime and also in the superconductor.

THERMODYNAMIC/ELECTRONIC CONTRIBUTIONS TO THE SUPERCONDUCTOR ENERGY GAP

Gibbs free energy provides the foundation for understanding the possibilities for formation of an energy gap in a superconducting system. Gibbs free energy at constant pressure and temperature, $G(p,T)$, is the difference between enthalpy, H, and temperature times entropy, $-TS$. The two terms provide two potential avenues for altering the free energies of systems, like normal conductors or components like conducting electrons. If we focus on the conducting electrons, which are intimately associated with the energy gap in superconductors, their enthalpy and entropy are the controlling factors.

There are two potential sources of enthalpy changes for conducting electrons in metals. The first is electron bond formation, e.g., Mott transitions in conductors, see Chapter 7. In quantum mechanics, these enthalpy changes are temperature

---

[*] Cubic closest packing is also referred to as a face-centered cubic lattice structure. In physics this crystal structure is referred to as tetragonal.
[†] Elements in column 14 of the periodic table are known as the carbon family.

independent, as they are in thermodynamics. In current theory, chemical bond formation is a temperature independent process, mediated entirely by forces due to particle
fields. Bond formation includes enthalpy associated with processes including: interactions of charges, Coulombic attraction and repulsion; and magnetic interactions,
magnetic coupling, spin pairing, etc. It corresponds to the bonding enthalpy and does
not seem a proper candidate for formation of a temperature dependent energy gap in
superconductors. Enthalpy changes in chemical systems are also a consequence of
changes in electron enthalpy associated with changes in electron momentum, k. This
electron enthalpy can be used to power machines or create heat, but it is not specifically related to the temperature dependent energy gap in superconductors.

In contrast to enthalpy changes, entropy changes are intrinsically temperature
dependent. This is true in both the Gibbs free energy and in the phenomenology of
entropy. Entropy changes associated with the end of dissipative electron scattering in
the phase transition between the normal conductor and superconductor states are the
source of the temperature dependent energy gap in superconductors. The subject can
be considered using Landauer's principle [33a, 33b, 33c], which links information and
thermodynamics, as well as standard thermodynamic considerations. The conserva-
tive estimate given for the Landauer limit in the measurements reported by Lutz, et al.
[33c] is of the order of $\ln 2 \cdot kT$ per erasure event, or in this case, per scattered electron.
[See ref. 33c, figure 3 c.] This data provides a credible beginning for understanding the
details of the superconductor temperature dependent energy gap, at least at onset, T_c.

At the normal conductor superconductor phase transition the mechanism for thermal conductivity, and heat capacity fundamentally changes. On the superconductor
side of the transition all of the conducting electrons are carriers of heat capacity with
direct quantum mechanical coupling to the lattice dynamics. The Born Oppenheimer
approximation fails here, as does the Sommerfeld equation, see Chapter 6. On the
normal conductor side of the phase transition, the Born Oppenheimer approximation
is applicable. The dominant electron carriers of thermal information are electrons in
s basis wave functions that exchange thermal energy with the lattice through partial
wave scattering and Fermi contact, see Chapter 8.

Changes in electron entropy have not previously been considered in discussions
of electron energy. The second law of thermodynamics assures us that changes in the
information content carried by electrons will result in changes in their free energy. It
is easy to see that the superconducting state, which completely lacks dissipative
electron scattering, is a more ordered state than its normal conductor counterpart
that harbors electrons with scattering potential. Changes in electron entropy associated with the end of dissipative electron scattering at the normal conductor superconductor phase transition will inevitably result is changes in electron free energy, and
contribute to the energy gap.

Since both the H and S terms depend on the current, both will also have an indi-
rect temperature dependence. The ΔS term is a statistical term that depends on the
loss of information associated with dissipative scattering.

If the energy of a uniform applied magnetic field, E_H, proportional to H_c^2 [33d], is
equal to ΔE, the superconductor energy gap, there is sufficient magnetic energy applied to
excite the system to the normal conducting state. The condition $E_H = \Delta E$ enables us to cal-
culate $H_c(T)$. Application of the standard treatment for critical phenomena [4,36] leads to

$$\Delta E = a\left(1 - \frac{T}{T_C}^{\,n} f(T)\right) \tag{11.2}$$

where n is a critical exponent, a is a proportionality constant, and ΔE corresponds to the energy gap of the superconductor. The first factor in Equation (11.2) corresponds to the threshold behavior near the critical point and $f(T)$ is a slowly varying function normalized to 1 at $T = 0$. We approximate $f(T)$ by the first-order McLaurin expansion,

$$f(T) = 1 + b\frac{T}{T_C} \tag{11.3}$$

Since the energy of the uniform magnetic field is proportional to H^2, from Equation (11.2) we find,

$$H_C(T) = H_C(0)\left(1 + b\frac{T}{T_C}\right)^{\!\frac{1}{2}}\left(1 - \frac{T}{T_C}\right)^{\!\frac{n}{2}} \tag{11.4}$$

A close approximation to this is

$$H_C(T) = H_C(0)\left(1 + \frac{b}{2}\frac{T}{T_C}\right)\left(1 - \frac{T}{T_C}\right)^{\!\frac{n}{2}} \tag{11.5}$$

High quality critical magnetic field data for indium and tin were published in 1960 by Shaw et al. [37]. We have conducted a preliminary evaluation of Equation (11.5) using this data with the results shown in Figures 11.5 and 11.6 and Table 11.3.

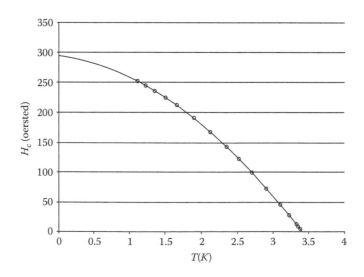

FIGURE 11.5 H_c v. T data (points) for indium, In (from R.W. Shaw et al. *Phys. Rev.*, 1960, *120*, 88–91), with fitted values, (solid line), from Equation (11.5).

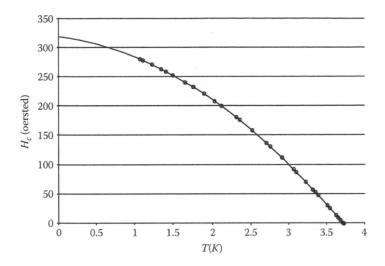

FIGURE 11.6 H_c v. T data (points) for tin, Sn (from R.W. Shaw et al. *Phys. Rev.*, 1960, *120*, 88–91), with fitted values (solid line) from Equation (11.5).

TABLE 11.3
Similarities between Indium and Tin

Element	$H_c(0)$	b	n	T_c	χ^2	COD
Indium	294.5	1.711	2.030	3.408	0.5965	0.999995
Tin	317.2	1.752	2.061	3.726	1.336	0.999996

Note: Data came from [37], fitting obtained by use of Psi Plot®; $H_c(0)$, b, and n values were obtained by regression using Equation (11.5). T_c values came from the data source [37]. χ^2 is the Chi-square statistic and COD is the coefficient of determination statistic.

Indium and tin are adjacent to each other in the periodic table. They are both robust superconductors and main group metals. Their critical magnetic field curves are very similar in shape. In both cases the fit to the simple thermodynamic model in Equation (11.5) is quite close (see Table 11.3).

Equation (11.5), because of its use of the critical exponent, has the potential to successfully deal with the variety of critical magnetic curves that are presented by elemental type I superconductors. It is reasonable to expect that new investigations in this area will prove fruitful.

EMPIRICAL/THEORETICAL REQUIREMENTS FOR SUPERCONDUCTIVITY

The material covered to this point can be summarized as a set of empirical/theoretical requirements for observation of superconductivity in a range of materials. These requirements apply to both type I and type II superconductors. The subtle differences between the two types of superconductors are discussed in the following chapter.

All superconductors can demonstrate a time independent magnetic field and corresponding current of substantial duration and no power source.

This is our best single empirical/theoretical definition of the superconducting state. It implicitly includes within the definition all of the individual requirements that must be met to have a superconducting magnet: (1) phase transition at a temperature specific to the material; (2) superconductor volume sufficient to allow expulsion of an applied magnetic field from any direction; (3) a temperature-dependent energy gap between the highest energy electrons in the superconductor and the lowest energy vacant orbitals of the system and energy gap under the same conditions, that is, a temperature dependent critical magnetic field; (4) zero electron scattering in the active supercurrent (this, in two dimensions, is the quantum Hall regime); and (5) bulk magnetic susceptibility of –1, perfect diamagnet: this is one of the manifestations of the Meissner–Ochsenfeld effect [35].

SUPERCONDUCTING PHASE TRANSITION AT A TEMPERATURE DETERMINED BY ZERO SCATTERING FOR THE HIGHEST ENERGY CONDUCTING ELECTRONS

From the original observation of superconductivity [1], superconductivity in a given material at one bar has been characterized by a dramatic drop of resistivity to zero in a conductor at a temperature that is characteristic of the material being examined. Sommerfeld's relationship points to electronic quantum mechanics as the source of resistivity in metals. Partial wave scattering is an established phenomenon in atomic physics that presents a structure for understanding resistivity (electron scattering) in metals as a function of temperature. It also provides the crucial distinction between pure elements that are superconductors and those that are not. This distinction, based on the presence or absence of s-basis wave functions in the conduction band for a metallic element, has the potential for development of a detailed understanding of T_c in superconducting elements and compounds. Such a detailed understanding is essential for the development of a comprehensive theory of superconductivity.

THE MATERIAL MUST BE THREE DIMENSIONAL AND LARGE ENOUGH TO SUPPORT A NONSUPERCONDUCTING BOUNDARY LAYER

Penetration layers for superconductors are essential for the development of the Meissner–Ochsenfeld effect [35], the expulsion of applied external magnetic fields that do not exceed the critical magnetic field.

Two-dimensional conductors that have zero electron scattering and zero electrical resistance cannot be superconductors because they cannot support a magnetic susceptibility of –1. There are now many examples of two-dimensional conductors that show zero electrical resistance. Graphene may be the best known case. Most such systems have been examined in quantum Hall experiments. A good example of this is the experiments performed by Božović, Bollinger, and coworkers on electric field switching of a Mott phase transition in a conducting cuprate that demonstrated zero electrical resistance [34]. In this experiment, it was shown that the prepared wafers were zero resistance conductors at the experimental temperature, but the two-dimensional system was not reported as having a superconductor volume magnetic susceptibility. The wafers had been designed specifically for measurement of magnetic susceptibility.

Wafers used in the experiments reported by Božović, Bollinge et al. were produced by epitaxial deposition and were of the order of two unit cells thick. Molecular composition of the wafers was $La_{2-x}Sr_xCuO_4$. This material is a high-T_c superconductor when operating as a bulk phase at appropriate conditions. The reported experimental results showed that there was no resistance to conduction in the system on the conductor side of the Mott transition. The transformation, too, must have been a standard Mott spin inversion. There was no evidence of the lattice structural changes that would have been essential had there been a Mott transition involving two phases that were both fully electron-pair bonded. The normal conductor electronic structure of the wafers in this experiment can possibly be settled by use of soft x-ray magnetic circular dichroism studies. Demonstration that an unpaired electron conductor can operate under specific conditions with zero electrical resistance will constitute a major contribution to our understanding of the theory of superconductivity [34].

The requirement of a three-dimensional conducting structure for superconductivity has been known since about the time of the Meissner–Ochsenfeld experiments, (~1933). My (RD) realization of this requirement came in a SQUID experiment with the acenaphthenyl radical, $C_{13}H_9 \cdot$.

acenaphthenyl radical, $C_{13}H_9$

This tricyclic odd-alternant hydrocarbon radical supports a ring current in a magnetic field. When it is maintained at 4.2 K, a temperature that assures no resistive scattering in the ring current conduction, the ring current is not persistent like a supercurrent because the volume magnetic susceptibility of the material is of the order of -10^{-5}, rather than -1.

A T DEPENDENT ENERGY GAP MUST EXIST BETWEEN THE HIGHEST OCCUPIED CONDUCTING AND LOWEST VACANT BANDS OF THE SUPERCONDUCTOR

Normal metals do not form an energy gap between occupied and vacant orbitals in the conduction band. Superconductors always have such a gap, which is the result of the change in conducting electron free energy associated with a decrease in entropy associated with loss of dissipative electron scattering in the normal conductor to superconductor transition. Mathematically, these two effects result in a thermo-dynamic change in the chemical potential of electrons in the conduction band of the superconducting system and form a temperature-dependent energy gap. For a mathematical description of the behavior of the critical magnetic field versus T, see Equation (11.5).

THE CONDUCTION BAND FOR THE SUPERCONDUCTOR MUST HAVE ZERO ELECTRON SCATTERING

Kamerlingh Onnes' 1911 discovery of zero resistivity in mercury metal at 4.15 K [1] launched one of the hallmarks of superconductivity—electrical conduction with-out electron scattering resistance. This feature of the phenomena is essential for the existence of persistent magnetic fields and currents. The large body of evidence presented in the previous chapters of this book consistently points to resistivity in metals as an electronic property of the system. Quanta of electrical resistance and magnetoresistance provide direct evidence on the point of the electronic origins of electron scattering resistance. Partial wave scattering meets the requirements for an electronic origin of resistance in metals. Partial wave scattering is the key to under-standing the pattern of elemental superconductors in the periodic table. Other than partial wave scattering, there is no current explanation for even the grossest features of the periodic table superconductivity patterns. Partial wave scattering sheds light on the fine details of the superconductivity patterns in the periodic table.

VOLUME MAGNETIC SUSCEPTIBILITY MUST BE –1

A modern reading of Meissner and Ochsenfeld's 1933 observations concerning superconductors [35] explicitly points to the requirement of a bulk magnetic suscep-tibility of –1 for a superconductor. The modern reading is only needed for the sake of SI units. Ochsenfeld was the experimentalist who first observed the expulsion of the earth's magnetic field from tin and lead cylinders as they were cooled below T_c. This happens because of the volume magnetic susceptibility being –1.

Bulk susceptibility of –1 and infinite three-dimensional conductivity* are the two factors that uniquely identify superconductors. These factors are inextricably inter-connected. When one appears, the other also appears. In the absence of these factors it is possible to observe electrical conduction with zero electron scattering resistance along a line or in a plane. It is also possible to observe a temperature-dependent energy gap in a two-dimensional system. Identification of the details of the high-order phase transition involved has not yet been accomplished to our knowledge. None of the quantum Hall regime or lower dimensional systems have yet been dem-onstrated to be capable of maintaining a current-generated magnetic field that is persistent without external power over a period greater than a year. Functionally, the difference between three-dimensional superconducting systems and zero resistance systems of two or fewer dimensions is that the superconducting systems cannot be

---

* Infinite three–dimensional conductivity is equivalent to zero electrical resistivity in three dimensions. One is the reciprocal of the other.

perturbed by external magnetic fields with field strength below the critical field. Lower dimensional, zero-resistance systems can be perturbed by such fields because their magnetic susceptibility is not at the extreme of –1. Bulk magnetic susceptibility does not exist for one- or two-dimensional systems, so this situation does not seem likely to change.

The Meissner–Ochsenfeld effect [35] is the source of –1 magnetic susceptibility in superconductors.* On the normal metal side of the phase transition line of critical points, the same condition is met by the cessation of electron scattering from the conduction band. In Chapter 9, we pointed out the relationship between electronic specific heat in superconducting elements, changes in electron scattering, and changes in magnetic susceptibility. These observations suggested that electron scattering in the conducting band is the source of the information transfer concerning local spin response to external magnetic perturbations that is essential for establishing bulk magnetic susceptibility for a normal metal. When the information transfer mechanism goes to zero at the superconductivity phase transition, the magnetic domain of the metal becomes the bulk volume of the superconductor. It is as if there were no spins to report. It is not that the spins are not there. The reporting of spin response from one part of the domain to another has ceased. At the same line of critical points, the Meissner–Ochsenfeld effect expulsion of subcritical magnetic fields defines the bulk susceptibility for the entire superconductor as –1.

The five empirical/theoretical requirements cited are the foundation for the defining requirement for superconductivity: persistent, current-generated magnetic field with no power source and substantial duration. Wires that carry zero resistivity currents may or may not be superconducting. The simplest test to determine superconductivity for a wire is a measurement of magnetic susceptibility under conditions where currents flow without resistance.

PERIODIC TABLE OF ONE-BAR SUPERCONDUCTORS: DOES IT FIT THESE REQUIREMENTS FOR SUPERCONDUCTIVITY?

Beryllium, Be, is the only one-bar superconductor in the periodic table that has an s ground state conducting band. Beryllium's T_c is 26 mK. It is not a robust superconductor. The electronic band structure for beryllium is known, and provides an explanation for its potential as a superconductor (see Figure 11.7) [38].†

Figure 11.7 shows that the $2p$-basis conducting bands are at the top of the conducting band, and more than 2 eV above the $2s$-basis bands in energy [34]. The experimental results for superconductivity at one atmosphere show that it is essential to be in the low millikelvin range of temperature ($T_c = 26$ mK) to find a state with zero contribution to conduction from $2s$-basis conduction bands. If $2s$ basis conduction bands were present, they would cause resistive electron scattering in electrical conduction at the lowest temperatures attainable by the mechanism of Fermi contact, partial wave scattering. From looking at Figure 11.7, it seems remarkable that a

---

* See the discussion in Thermodynamic Requirements for a Superconductor Energy Gap.

† This figure was previously seen as Figure 8.2.

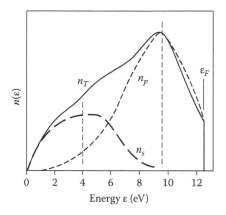

Energy ε (eV)

FIGURE 11.7 Density of states for clean beryllium (from R.W. Shaw et al. *Phys. Rev.*, 1960, *120*, 88–91). n_T, calculated electronic density of states (from T. L. Loucks, and P. H. Cutler, *Phys. Rev.*, 1964, *133*, A819–29). n_p, partial density of states for electrons with *p* symmetry, obtained from x-ray emission studies (from T. Sagawa, *Soft X-Ray Spectrometry and the Band Structure of Metals and Alloys*, D. J. Fabian, Ed., 1968, Academic, New York; p. 29).

temperature of 26 milikelvin is required to reach the asymptotic zero of the number density of the 2s-basis conduction bands for beryllium.

Noble metals, column 11 in Table 11.1, and alkali metals, column 1 in Table 11.1, both have ns^1 electronic ground states as atoms, and ns^1 basis ground state conducting bands. Neither column 1 nor column 11 shows any one-bar superconductors, although the noble metals are the best normal conductors in the periodic table. Lithium, Li, and cesium, Cs, column 1, both have isolated electronic states at high pressures, 50 and 8 GPa [5] respectively, that are superconducting. There are no known conditions that render pure copper, silver, or gold, superconducting.

The block of one-bar superconducting elements between lanthanum, La, and tin, Sn, shaded elements between columns 3 and 14, can all be understood as np^x superconductors. In the metal ground state, the superscript "x" varies from 1 to 2 for this group of elements. In the early and middle transition metals, columns 3 and 4, and 5–7, the relative values of the critical temperatures suggest that the odd numbered columns (3, 5, 7) have the most favorable electronic structures for superconductivity. Manganese, Mn, has the highest resistivity of any element. This suggestion of bond localization in manganese is related to its lack of superconductivity.

Superconducting lanthanum should have a $6p^1$ basis function conduction band. The measured number of carriers for La, using a single band model at room temperature, was 2.9 [41]. The next lower-energy conductor configuration would be either $[Xe]6s^15d^16p^1$, which would have two carriers per atom, or $[Xe]6s^25d^06p^1$, which has only one. For superconducting lanthanum we believe that the latter configuration is the most probable.* This can be checked by experiments in the quantum Hall

---

* The $[Xe]6s^25d^06p^1$ configuration is congruent with robust superconductivity. A configuration with an open 6s subshell would only be consistent with a weak superconductor.

regime with a film of elemental lanthanum at temperatures below the 6 K critical temperature for La.

Vanadium family superconductors have an $ns^2(n-1)d^1np^2$ basis conduction band and, like lanthanum, a closed ns^2 subshell. Niobium, in this family has the highest one-bar critical temperature for the elements. This maximum T_c and the corresponding maximum value for the ~0 K H_c may justify the assignment of two carriers per atom for the vanadium family. Once again, quantum Hall regime experiments with these elements will provide direct evidence concerning the number of carriers in the superconducting state.

Continuing our focus on the robust superconductors ($T_c > 1.5$ K) in the transition metals, we move to column 7 in the table, with technetium, Tc, and rhenium, Re. By analogy with manganese [42], Tc and Re should have four carriers at ambient temperature. In the superconducting state, their atomic electronic configuration should be [X] $ns^2(n-1)d^4np^1$,[*] and the superconducting band should have an np^1 basis, with only one carrier per atom using orbitals formed from the atomic ground state configuration. Data from beryllium suggests that the existence of an open ns subshell, even with a large s–p energy gap, will drive the T_c for the superconductor to very low temperatures. Relatively high values for T_c for both technetium and rhenium are consistent with our expectation that the highest energy superconducting band for these elements has an np^1 basis, and the ns subshell for the lattice atoms is doubly occupied and closed.

Expectation of a reduced number of carriers in the superconductor compared to the normal conductor extends to the transition metal superconductors in column 12. In the zinc family, the number of carriers at room temperature is two; however, for the superconductor, the atomic electronic configuration should be $[X]ns^2(n-1)$ d^9np^1. Zinc family superconductors are thus expected to have one carrier per atom, since the overlaps for the $(n-1)d$ atomic basis functions with those of adjacent atoms are much too small for them to participate in a conduction band. It is conceivable that both Zn and Cd have open s subshells as superconductors. This would account for their relatively low critical temperatures.

Requirement of an np^1 basis conduction band for the elements in column 13— aluminum, gallium, indium, and thallium, has yet to be verified by experiment. The same is true for the superconducting conduction band for group 14, tin and lead which must be np^2. In all of these cases, experiments must be conducted in the quantum Hall regime at temperatures below T_c, with or without magnetic fields above H_c for the superconductors, depending on sample thickness.

Having looked at the robust superconductors, we need to consider the elements in the adjacent columns. Group 4 of the periodic table has three weak elemental one-bar superconductors. All of the critical temperatures for titanium, Ti, zirconium, Zr and hafnium, Hf, are in the narrow range between 1.5 and 0.1 K. In the superconducting state, their atomic electronic configuration should be $[X]ns^2(n-1)d^1np^1$, and the superconducting band should have an np^1 basis, with only one carrier per atom. Explanations for the low critical temperatures compared to the vanadium family in column 5 involve an emerging population of conducting states containing ns^1, singly

---

[*] [X] in this atomic electronic configuration represents the immediately preceding rare gas electronic structure from column 18 of the periodic table.

occupied s-wave functions, when thermal energies are large enough. One state that could be responsible for this emergence is the single carrier state, $[X]ns^1np^3$. The corresponding state with a half-filled p subshell in column 5 would have a closed, doubly occupied s subshell, which would be a nonconductor. Half-filled subshells are particularly stable and are also nonconductors.

Molybdenum, Mo, a weak elemental superconductor, is the only one-bar super-conductor in group 6 of Table 11.1. The atomic state supporting superconductivity is $[Kr]5s^24d^35p^1$. By analogy with the titanium family, the emerging partially occupied s state that would suppress the critical temperature for molybdenum could be $[Kr]5s^14d^25p^3$, $[Kr]5s^14d^35p^2$, or $[Kr]5s^14d^45p^1$.

As we proceed to the right in the transition metal series, each new column adds an $(n-1)d$ electron. This increase in the electronic screening of the nuclear charge from the outer s electrons raises the energy of the outermost s level relative to the corresponding p level. In the metals, where the conduction bands come from the valence level, this means that s electron bands are more likely on the right half of the transition series than they are on the left. It is not a surprise, then, that columns 8 through 11 have only three one-bar superconductors, all of them weak or very weak.

The screening effect of adding d electrons as we move to the right in the periodic table also provides a direct explanation for the observed order of the critical temperatures for the robust superconductors in period 6 in the d series: La, 6.0; Ta, 4.4; Re, 1.7 K, respectively.

To complete this second tour of the periodic table of superconductors, we need to look again at the copper family, group 11. Silver is the best ambient temperature conductor of the elements. Copper is second best, and gold is close behind. The electronic ground state for these elements is $[X]ns^1(n-1)d^{10}$. These elements are intrinsic nonsuperconductors, though they are known to participate in numerous intermetallic superconducting compounds. For an early introduction to this field, see an excellent paper by Matthias [43]. Intermetallic gold–diindium, $AuIn_2$, has received relatively recent attention as a very weak superconductor [44]. Specific intermetallic bond formation, in this case, suppresses the availability of the half occupied gold $6s$ conduction band, so that a critical temperature of 37 μK was observed. This system also exhibits interplay between nuclear magnetism and superconductivity. One-bar critical temperature for indium is 3.4 K, roughly five orders of magnitude higher than the T_c of $AuIn_2$.

Resistivity for pure copper as a function of temperature is instructive in many ways. Figure 11.8 shows resistivity data collected for a sample of highly refined copper prepared in the 1950s. The Copper Development Association [45] presents contemporary data for refined copper that shows relative resistivity ratios, $\rho(273\ K)/\rho(4\ K)$, that vary from 10 to 2000. For the data in Figure 11.8, the relative resistivity ratio is approximately 500. At a relative resistivity ratio of 2000, the highest purity copper currently commercially available, the residual resistivity of copper below 10 K is approximately 7.5×10^{-12} Ω•m [46]. It is worthy of note that as the relative resistivity ratio changes by a factor of 200, the curve shape for copper resistivity versus temperature is essentially constant. Low temperature resistivity for copper with a relative resistivity ratio of 2000 approaches the asymptotic resistivity of ultra pure

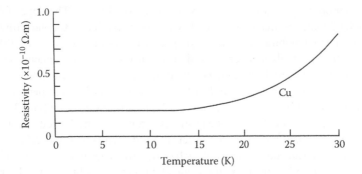

FIGURE 11.8 Resistivity of refined copper at low temperature.

copper caused by partial wave scattering in a 4s-basis conduction band. Previously, it was thought that the residual resistivity of copper was entirely due to impurities and lattice defects, and that a pure, defect-free single crystal would show no resistance at a low enough temperature. Copper's $[Ar]4s^13d^{10}$ ground state has Fermi contact partial wave scattering at the lowest attainable temperatures, so copper will always have a residual resistivity, even in the micro-kelvin range. This may be clearer using an example taken from graphene.

Figure 11.9 illustrates the ρ_{xx} resistance measurements obtained, in conjunction with a Hall effect experiment, for a sample of bilayer graphene in the temperature range from ~0 to 300 K [47].

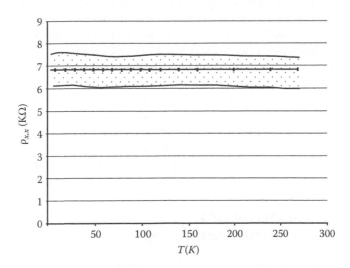

FIGURE 11.9 Hall ρ_{xx} resistance of bilayer graphene Hall probe v. T. The upper and lower line in the band in the figure are the error bars on the reported measurements. The straight line in the middle of the band is the average of the reported measurements with the locations of the temperatures for the reported measurements obtained from the original figure. (From K. S. Novoselov et al., *Nature Phys.*, 2006, 2, 177–80; Figure 2, b, c.)

Molecular orbitals for graphene-like systems are well known. The slope of the resistance versus T indicates that bilayer graphene as prepared has a $3s$ conducting band.

Reported carrier mobilities for the bilayer graphene devices were around $3000\,cm^2V^{-1}s^{-1}$ [47], an unexpectedly smaller value than usually seen with graphene. The apparent decrease in carrier mobility may be due to the presence of switching electric field polarity caused by biexcitonic Bose–Einstein-correlates [49,50]. Temperatures ranged from mK to approximately the ice point for the plot in Figure 11.9 [48]. To preserve electrical neutrality in the two-graphene sheets, excitations are biexcitonic, producing a delocalized π radical cation in one sheet and a delocalized π radical anion in the adjacent sheet, with an equal number of each of the two types in both sheets. The z-axis electric field produced by the biexcitonic correlates is responsible for the observed chirality of the electrons in this unique quantum Hall effect.

The fact that the graphene resistance in Figure 11.9 is essentially constant at approximately 7. $k\Omega$ (author accuracy of measurement in Figure 11.7, $\pm 10\%$) [47], is not trivial to explain. For the experimental temperature range, ~273 K, the observed constant resistance would be difficult to explain with any model other than partial wave scattering from carbon $3s$-basis wave functions. For copper using a $4s^1$ conduction band, the temperature at which the resistivity becomes constant* is ~8 K, (Figure 11.8). At a temperature between ~8 and ~10 K in ultra-pure copper conducting bands using $4p^1$ basis, functions start to be thermally populated and copper resistivity undergoes an exponential increase with temperature (see Equations [8.13] and [8.14]). In carbon using a $3s^1$-based conduction band, the temperature range for constant resistivity appears to be above 270 K. This is due to the large energy gaps between potential conduction bands in graphene. To understand this behavior, it is necessary to look at the equations for partial wave scattering, (Equations [8.5] to [8.14]). In the approximations that we used for the threshold orbital-angular-momentum-dependent scattering cross section, Equation (8.13), the cross section is given by a collection of constants times the wave vector to the power $4l$. For s electrons, $l = 0$ and the scattering cross section at threshold is a constant with temperature as shown in Figure 11.8 below 8 K and in Figure 11.9 below approximately 273 K.

Low-temperature resistance data for copper, Figure 11.8, and graphene, Figure 11.9, provide what is probably the strongest available evidence that partial wave scattering is the operational process of resistive electron scattering in metallic conductors. It would require the development of as yet unknown processes if phonon-based electron scattering were to provide the foundation for understanding these resistance observations.

ESTIMATING THE ORDER OF T_C FOR ELEMENTAL SUPERCONDUCTORS

In our discussion of partial wave scattering in Chapter 8, we presented the data in Table 8.2, which is reproduced here as Table 11.4. In the model we used for partial wave scattering at threshold, the scattering cross section for s-basis functions is a constant

---

* This is the temperature below which wave functions with $l > 0$ are no longer a part of the conduction band.

TABLE 11.4

Order of magnitude resistivity ratios versus *T*

Ratio $\diagdown$ T	300 K	200 K	100 K	10 K	1 K
p/s	$3*10^{-4}$	$1*10^{-4}$	$3*10^{-5}$	$3*10^{-7}$	$3*10^{-9}$
d/s	$5*10^{-8}$	$1*10^{-8}$	$7*10^{-10}$	$7*10^{-14}$	$7*10^{-18}$
f/s	$8*10^{-12}$	$7*10^{-13}$	$1*10^{-14}$	$1*10^{-20}$	$1*10^{-26}$

as a function of temperature because *l* in the temperature exponent is zero. By using ratios of the cross sections for *p*, *d*, or *f* basis functions to those of the corresponding *s*-basis functions, the unknown constants are eliminated. The ratios for threshold scattering cross sections for different atomic basis functions as a function of temperature are presented in Table 11.4. Equations and assumptions are given in Chapter 8 in the section Molecular Orbital Approaches to Resistivity. Most theory serves the function of developing an understanding of experimental facts. Certainly, there are occasions where *de novo* predictions are made, for example, Landau's prediction of the superfluidity of ^{3}He [51]. Most theory, however, serves the function of improving our ability to understand and categorize known experimental facts.

Table 11.4, along with the approach of the previous section, can be a starting point for developing an understanding of the critical temperatures of the elemental superconductors. To use the table in this way requires that we employ a semiempirical approach to the problem of T_c prediction. Our current theoretical understanding is that all of the known elemental one-bar superconductors have *p*-basis conduction bands when they are in a superconducting phase. This understanding is derived from the analysis of the pattern of one-bar superconductors in the periodic table that has been discussed throughout this book, starting in Chapter 2.

Superconductors with *d*-basis conducting bands are well known in the high T_c superconductors. Direct comparisons between elemental superconductors and perovskite superconductors are not reasonable because the perovskites all have an oxygen ion between every pair of copper ions in the conducting plane. The presence of the oxygen ions and the relatively long copper–copper bond lengths* in these systems are responsible for the weak coupling between the copper *d* atomic basis functions on different atoms that characterizes the conduction band of the system. Type II superconductors are the subject of Chapter 12, where the critical temperatures, among other things, are discussed. Type II superconductors that utilize *p*-basis function conduction bands include niobium, technetium, vanadium, and a number of intermetallic compounds.

Robust elemental superconductors at one bar have critical temperatures that vary from 9.25 to 1.5 K. Niobium sets the upper bound. The lower bound is an arbitrary selection. Using the values in Table 11.3, a T_c range of 9.25 to 1.5 K would cover a *p/s* resistivity ratio of approximately $2 \bullet 10^{-7}$ to $4 \bullet 10^{-9}$. Our problem of understanding the relative values of T_c for a series of elements would be a lot easier if we could just focus on the energy of the band that becomes superconducting. Discussion of the pattern of superconductors in the periodic table presented above indicates that we must

---

* The long copper bond lengths in perovskites result in part from the formal 2+ charge on each copper.

also be critically aware of the energies of the nonsuperconducting wave functions in the system. The example of beryllium shows that it is possible for an element to be a superconductor, even if s-basis and p-basis conduction bands are operative, if one goes to a low enough temperature. The Be one-bar T_c is 26 mK.

Understanding the value of T_c for an elemental superconductor requires knowledge of the density of electronic states for the metal as a function of temperature. Once the density of states as a function of temperature is known, it will be necessary to determine the temperature at which the density of s-basis wave functions in the conduction band is approximately zero. Using the approach of Table 11.4, the value of the p/s ratio should be approximately the same for all elements in the same period. If T_c for an element is roughly 10 K, when the ratio of the density of p states to s states reaches roughly 10^{-7} we should be near the superconductivity transition. You can think of this as a signal to noise problem with the value in Table 11.4 as the noise level that nominally corresponds to zero. If at the threshold temperature, there are no lower energy s-basis conduction orbitals, and there exists a nonscattering conduction orbital at the same energy, the phase transition to superconductivity will occur. When this happens, all of the current will drop into the nonscattering orbitals, and the processes associated with the superconducting phase transition will occur.

Fundamental requirements for a one-bar superconducting element are:

1. The element must be a conductor at low temperature.
2. The Fermi level conduction band must have only $l > 0$ basis functions.
3. For robust elemental superconductors, the valence level s subshell is closed.
4. For weak and very weak superconductors, the valence level s subshell is open but well separated in energy from higher l-basis conduction orbitals.

Type II superconductors have considerably more magnetic domain structure than their Type I counterparts. Increasing magnetic structural complexity and maximum attainable field strength are intimately related in superconducting magnets. Because of this, type II superconductors are dominant in the arena of commercial superconductors. Type II superconductors are the subject of the next chapter.

REFERENCES

1. H. Kamerlingh Onnes, *Leiden Comm.*, 1911, 120b, 122b, 124c.
2. See: P. F. Dahl, *Superconductivity*, 1992, American Institute of Physics, New York.
3. D. Shoenberg, *Superconductivity, second edition with Appendix*, 1960, University Press, Cambridge; Fig. 77 a, b, pp. 224–225.
4. See, e.g., (a) H. E. Stanley, *Introduction to Phase Transitions and Critical Phenomena*, 1971, Oxford University Press, New York; (b) J. M. Yeomans, *Statistical Mechanics of Phase Transitions*, 1993, Oxford University Press, New York.
5. C. Buzea, and K. Robbie, *Supercond. Sci. Techno*l., 2005, *18*, R1–R8.
6. R. W. Shaw, D. E. Mapother, and C. Hopkins, *Phys. Rev.,* 1960, *120*, 88–91.
7. B. B. Goodman, and E. B. Mendoza, *Phil. Mag.*, 1951, *42*, 594–602.
8. A. D. Misener, *Proc. Roy. Soc. A*, 1940, *174*, 262–72.
9. R. A. French, J. Lowell, *Phys. Rev.*, 1968, *173*, 504–9.
10. B. B. Goodman, *Nature,* 1951, *167*, 111.
11. J. G. Daunt, and K. Mendelssohn , *Proc. Roy, Soc. London A*, 1937, *160*, 127–36.

12. J. M. Lock, A. B. Pippard, and D. Shoenberg, *Proc. Camb. Phil. Soc.*, 1951, *47*, 811.
13. J. G. Daunt, and K. Mendelssohn, *Proc. Roy. Soc. A*, 1937, *160*, 127–36.
14. J. G. Daunt, and C. V. Heer, *Phys. Rev.*, 1949, *76*, 715–7.
15. A. Wexler, and C. S. Corak, *Phys. Rev.*, 1950, *79*, 737–8.
16. N. Kürti, and F. E. Simon, *Proc. Roy. Soc. A*, 1935, *151*, 610–23.
17. L. Y. L. Shen, N. M. Senozan, and N. E. Phillips, *Phys. Rev. Letters*, 1965, *14*, 1025–7.
18. R. Radebaugh, and P. H. Keesom, *Phys. Rev.m* 1966, *149*, 209–16.
19. A. T. Hirshfeld, H. A. Leupold, and H. A. Boorse, *Phys. Rev.*, 1962, *127*, 1501–7.
20. R. J. Trainor, and M. B. Brodsky, *Phys. Rev. B*, 1975, *12*, 4867–9.
21. W. H. Keesom, and J. A. Kok, *Akademie der Wetenschappen, Amsterdam, Proceedings*, 1932, *35*, 743–48.
22. W. S. Corak, and C. B. Satterthwaite, *Phys. Rev.* 1956, *102*, 662–6.
23. C. A. Bryant, and P. H. Keesom, *Phys. Rev.*, 1961, *123*, 491–9.
24. W. H. Keesom, and J. A. Kok, *Physica*, 1934, *1*, 503–12.
25. A.Yu. Sadykova, A. S. Saykaeva, A.V. Kostochko, A. N. Glebov, and V. G. Moozyukov, *Int. J. Quantum Chem.*, 1992, *44*, 935–47.
26. F. London, *Phys. Rev.*, 1948, *74(5)*, 562–73.
27. N. F. Mott, *Proc. Phys. Soc. (London) A*, 1949, *62*, 416–22.
28. L. Pauling, *The Nature of the Chemical Bond*, 1939, Cornell University Press, Ithaca.
29. G. N. Lewis, *J. Amer. Chem. Soc.*, 1916, *38*, 762–85.
30. See, e.g., http://www.gaussian.com/.
31. See, e.g., http://www.schrodinger.com.
32. See, e.g., F. A. Cotton, and X. Feng, *J. Amer. Chem. Soc.,* 1997, *119*, 7514–20.
33a. R. Landauer, Irreversibility and heat generation in the computing process. IBM *J. Res. Develop.*,1961, 5, 183–191; b) R. Landauer, *Nature*, 1988, 335, 779–784; c) A. Bérut, A. Arakelyan, A. Petrosyan, S. Ciliberto, R. Dillenschneider, E. Lutz, *Nature*, 2012, 483, 187-189; d) . J. D. Jackson
33. J. D. Jackson, *Classical Electrodynamics*, third edition, 1998, John Wiley, New York; p. 213.
34. A. T. Bollinger, G. Dubuis,, J. Yoon, D. Pavuna, J. Misewich, and I. Božović, *Nature*, 2011, *472*, 458–460.
35. W. Meissner, and R. Ochsenfeld, *Naturwissenschaften*, 1933, *23*, 787–8.
36. M. Marder, *Condensed Matter Physics*, second edition, 2010, John Wiley & Sons, New York; Critical Phenomena, pp. 743–754.
37. R. W. Shaw, D. E. Mapother, and D. C. Hopkins, *Phys. Rev.*, 1960, *120*, 88–91.
38. R. G. Musket, and R. J. Fortner, *Phys. Rev. Letters*, 1971, *26(2)*, 80–2.
39. T. L. Loucks, and P. H. Cutler, *Phys. Rev.*, 1964, *133*, A819–29.
40. T. Sagawa, *Soft X-Ray Spectrometry and the Band Structure of Metals and Alloys*, D. J. Fabian, Ed., 1968, Academic Press, New York; p. 29.
41. C. J. Kevane, S. Legvold, and F. H. Spedding, *Phys. Rev.*, 1953, *91*, 1372–9.
42. M. Marder, *Condensed Matter Physics*, second edition, 2010, John Wiley & Sons, New York; Table 17.1, p. 499.
43. B. T. Matthias, *Phys. Rev.*, 1953, *92,* 874–6.
44. S. Rehmann, T. Herrmannsdörfer, and F. Pobell, *Phys. Rev. Letters*, 1997, *78*, 1122–5.
45. http://www.copper.org/resources/properties/cryogenic/homepage.html.
46. Ibid.
47. K. S. Novoselov, E. Mccan, S. V. Morozov, V. I. Fal'ko, M. I. Katsnelson, U. Zeitler, D. Jiang, F. Schedin, and A. K. Geim, *Nature Phys.*, 2006, *2*, 177–80.
48. *Ibid.* Figure 2, b, c.
49. J. P. Eisenstein, and A. H. MacDonald, *Nature*, 2004, *432*, 691–4.
50. S. A. Moskalenko, and D. W. Snoke, *Bose–Einstein Condensation of Excitons and Biexcitons*, 2000, Cambridge U. Press, New York.
51. L. D. Landau, L. D., *Zh. Eksp. Teor. Fiz.*, 1956, *30*, 1058.

12 Type II Superconductivity

In 1950, Landau and Ginzburg published a paper on the theory of superconductivity that laid the foundation for development of a theory describing high T_c, high H_c superconductors [1,2]. Magnetic flux quantization was introduced in this paper, which provided the basis for the description of the electron vortex lattice that is observed in type II superconductors. In the Ginzburg–Landau theory, a critical parameter is the ratio of the penetration depth, λ, to the coherence length, ξ.

The structure shown in the magnetic field intensity for vanadium in Figure 12.1 is not known in type I superconductors. It is a consequence of the vortex lattice structure in vanadium as a type II superconductor. The second-order phase transition that leads to the formation of the vortex lattice appears as a result of a thermodynamic instability associated with the negative surface energy at sufficiently high magnetic field strengths in superconductors [6,7].

Abrikosov developed our understanding of formation of the vortex lattice as a response to the negative surface energy states that appear in type I superconductors at a sufficiently elevated magnetic field [3,4].

Elemental superconductors that exhibit type II behavior include vanadium, V [5,6]; niobium, Nb [6]; and technetium, Tc [7]. The report by Williamson [4] concerned the upper critical magnetic field, H_{c2}, for both vanadium and niobium. Data for H_{c2} for vanadium from this paper is presented in Figure 12.1. The upper critical field for a type II superconductor, H_{c2}, is the maximum magnetic field attainable at a given temperature and geometry for the superconductor vortex lattice (see the following text).

ISOTOPE EFFECTS IN SUPERCONDUCTORS

Early discoveries of isotope effects on critical temperatures in superconductivity [8,9] focused attention on the zero-point vibrations of the lattice in superconductors [10]. The proportional relationship between oscillator frequency and the square root of isotope mass was known prior to 1950. The relationship has the same mathematical basis as Hooke's law for a simple harmonic oscillator. In studies of superconductivity, isotope effects are often expressed in terms of an elemental mass exponent, α (Equation 12.1):

$$T_C \propto M^{-\alpha} \tag{12.1}$$

$$\alpha = -\frac{d \ln T_C}{d \ln M}$$

For type I superconductors, 0.5 is a more or less satisfactory value for the exponent α. This value follows Hooke's law. Type II superconductors have been found to have a wide range of elemental isotopic exponents. The M in the two versions of equation 12.1 is the mass involved in the harmonic oscillator approximation.

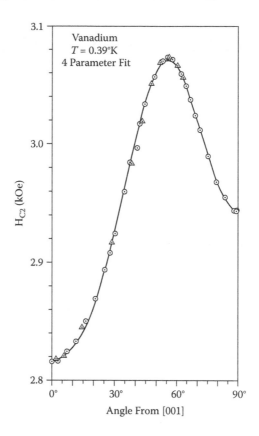

FIGURE 12.1 Orientation dependence of H_{c2} in the $(0\bar{1}0)$ plane for $t=0.072$. (Note: t is T/T_c. For vanadium, T_c is 5.4 K.). The solid line is a fit of the data using four cubic harmonics. (From S. J. Williamson, Phys. Rev. B, 1970, 2, 3545–56.)

The relatively recent discovery of the superconductivity of magnesium diboride, MgB_2 [11], has added new dimensions to isotope effect data for superconductors in the literature. Magnesium diboride has a T_c of 39 K for the natural abundance intermetallic compound. Critical temperatures for $Mg^{10}B_2$ and $Mg^{11}B_2{}^*$ were reported as 40.2 and 39.2 K [12], respectively. This 1 K difference in T_c for a change in formula mass of roughly 2 amu in 46 amu for $Mg^{11}B_2$ is larger than the harmonic oscillator model would predict. Before we get into the details of the isotope effects, we will comment briefly on the $2p$, $3p$-basis functions for the conduction bands in MgB_2.

THE CONDUCTION BAND IN MGB₂

Some investigators are inclined to think that all type II superconductors have the same type of conduction bands. This would be true if the designation of type II superconductors was based on a specific property of the conduction band. Designation of

---

* The isotopes used in these experiments were reported to be >99.5% pure isotopic composition as stated.[12]

type II superconductors is based on the structure and dynamics of the magnetic field in the superconductor. When the magnetic field in a superconductor is high enough to create a negative surface energy for the superconductor, a second-order phase transition occurs, and the magnetic field structure shifts to a vortex lattice [6,7]. This is true for MgB_2 and for the cuprate superconductors, whose isotope effects are discussed later in this chapter.

Cuprate superconductors have d-wave conduction bands and critical temperatures that vary from less than 40 to greater than 120 K. MgB_2 has only $2p$-basis functions from boron and $3p$-basis functions from magnesium with which superconducting bands can be constructed. The crystal structure for MgB_2 is known [13]. The crystal has a graphene-like plane of boron atoms with planes of magnesium ions on either side [11,13]. In this arrangement, the crystal is a three-dimensional conductor and attains a volume magnetic susceptibility of −1 at the superconducting critical temperature [11]. Using the scaling of Table 11.4, the ratio of p/s scattering at 40 K from our simple partial wave model is ~ $6•10^{-6}$, which suggests to us that this may be close to the highest observable T_c for a p-basis superconductor.

Isotope Effects in Type II Superconductors

Isotope studies of magnesium diboride, that included Mg isotopes showed the system to be more complex than restriction of the isotope effect to the boron layers in the crystal [14]. Figure 12.2 is plotted using data in the paper by Hinks et al. [15]. It illustrates the relative magnetization, M, for samples that contained ^{24}Mg, ^{n}Mg, ^{28}Mg, ^{10}B, and ^{11}B. ^{n}Mg is the natural isotope distribution for Mg, which is 78.99% ^{24}Mg; 10.00% ^{25}Mg; and 11.01% ^{26}Mg [16]. It is possible to see in Figure 12.2 that the influence of mass from magnesium on the shape of the magnetization curve at T_c is different from that seen for mass from boron. The authors point out that the low isotope coefficient for Mg, $\alpha = 0.32$ compared to 0.5, for the harmonic oscillator approximation, "could be due to complex materials properties." The value is quite close to the same coefficient obtained for B , α_B 10% $T_c = 0.31$ [15].

One can see from Figure 12.2 that the effect of adding mass to boron or magnesium is different with respect to the shape and position of the magnetization curves for the superconductor. Figure 12.2 shows that the magnetization curve for $^{26}Mg^{11}B_2$ is distinct from those of $^{n}Mg^{11}B_2$ and $^{24}Mg^{11}B_2$. The figure also shows that $^{n}Mg^{10}B_2$ has the highest T_c of the examined isotope combinations [15]. The original authors suggested that "this may be due to some small impurity content in the Mg isotopes..." [14]. All of these differences must be grounded in the chemical structure of the material, which is known [13]. The crystal structure of magnesium diboride has layers of boron atoms connected in a graphene-like structure. On either side of a boron layer there is a hexagonal closest packed layer of magnesium 2+ ions. See the original report of this high T_c intermetallic compound by Nagamatsu et al. for a drawing of the lattice components [11]. The bulk of the current in the conductor is carried by boron wave functions [17].

The detailed mass dependence of the magnetization curves for isotopomers of magnesium diboride illustrates the lack of applicability of the Born–Oppenheimer approximation to active superconductors. If the Born–Oppenheimer approximation

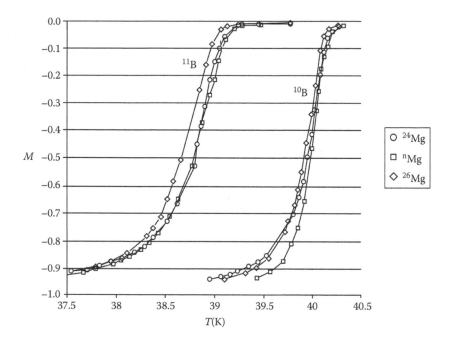

FIGURE 12.2 The superconducting transitions for the isotopically substituted magnesium diboride, MgB_2, samples. The two sets of curves present the relative magnetization for all samples. ^{n}Mg indicates samples with natural abundance Mg. ^{24}Mg and ^{25}Mg samples were prepared from isotopically enriched Mg at respectively 99.92% and 99.94%. ^{10}B and ^{11}B samples were synthesized from boron isotopically enriched at respectively 99.52% and 99.49%. The small Mg isotope effect on the ^{11}B, ^{10}B isotope effect that separates the two sets of curves is visible in the curve shapes for the respective samples. (From D. G. Hinks, H. Claus, and J. D. Jorgensen, *Nature*, 2001, *411*, 457–60. Data from, Figure 1, p. 458.)

were valid for this superconductor, we would expect the isotope exponents to be much closer to 0.5. A successful theoretical approach to this problem can be anticipated to include some form of vibrational electronic quantum mechanics.

Oxygen and other isotope effects including copper have been widely studied in high-temperature superconductors [18]. Table 12.1 presents information on the range of element mass exponents cataloged in Franck's coverage of the literature prior to 1994.

In Table 12.1, n is the number of measurements of the oxygen isotope mass exponent, α_O, for different materials, reported in the original table; α_O max is the maximum value of α_O in the original table; and the critical temperatures that follow α_O max or min are the critical temperatures for those specific measurements. Table 12.1 shows that a harmonic oscillator approach to the oxygen isotope effect problem in this group of superconductors is not likely to prove satisfactory. The critical temperatures associated with the measurement of the maximum and minimum isotope effect in each of Franck's original groupings show that, in general, higher values for T_c are associated with the minimum values of α_O for the group. Substantial variations

TABLE 12.1

Oxygen Mass Exponents for Isotope Effects on T_c of type II Superconductors

Formula	n	α_O max	T_c (K)	α_O min	T_c (K)	Ref. 18, Ref. #
$YBa_2Cu_3O_{7-\delta}$	31	0.07	91	0.0	91	217
$YBa_2Cu_3O_7$[†]	53	0.53	31	0.0	85	232
$YBa_2Cu_4O_8$[*]	16	0.38	53	0.02	89	236
$La_{2-x}M_xCuO_4$[‡]	44	0.93	28	0.06	34	245
$La_{2-x}M_xCuO_4$[§]	57	1.3	13	0.12	36	248

[†] Mono- and di-substituted on Y or Ba.
[*] Doped, substituted on Y or Cu.
[‡] M is Sr or Ba.
[§] M is Sr or Ca. Cu can have fractional substitution with Fe, Co, or Ni.

Source: J. P. Franck, in *Physical Properties of High Temperature Superconductors IV*, D. M. Ginsberg, Editor, 1994, World Scientific Publishing, Singapore; pp. 189–293.

in the maximum and minimum values of the oxygen mass exponent with chemical structure and conditions indicate that these factors will demand attention in any successful theory concerning these effects.

Oxygen isotope effects in a ferromagnetic cuprate, $RuSr_2GdCu_2O_8$, were reported by Pringle et al. in 1999 [19]. The oxygen mass exponent, α_O, for the superconducting critical temperature was 1.6. This is more than three times the exponent anticipated using a harmonic oscillator model. In contrast, the relative Raman frequency shifts for four oxygen Raman lines in the ^{16}O isotopomer of this superconductor produced values that were relatively close to the harmonic oscillator expectation, the straight line in Figure 12.3. The figure shows that the four individual lines are not very close to being parallel to the harmonic expectation.

Under conditions where the Born–Oppenheimer approximation is applicable, there is a substantial literature to suggest that the harmonic oscillator approximation closely characterizes the isotope effects associated with exchange of ^{18}O for ^{16}O illustrated of the general kind shown in Figure 12.3. The lack of a close fit between the experimental data and the harmonic theory suggests the need for a different theoretical treatment of this system.

The low-temperature heat capacity of superconducting $Mg^{11}B_2$ was reported less than a year after the isotope effect experiments on the same material [20]. The authors explain their presentation of the data with the following paragraph:

Below 2 K, there is an H-dependent hyperfine contribution to C. There are also several H-independent features in C, including an "upturn" below 2 K, that are probably associated with small amounts of impurity phases. Partly for that reason, most of the interpretation of the results is based on an analysis of the differences, C(H) – C(9 T), in which the H-independent extraneous contributions and most of the contributions of the addenda, including the varnish, cancel. C(H) – C(9 T) was calculated after the data were corrected for the hyperfine contributions and a small H-dependent part of the heat capacity of the sample holder. Since C is the sum of an H-dependent electron

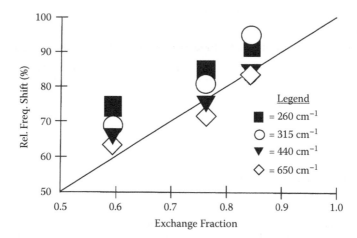

FIGURE 12.3 Illustrating the relative frequency shift for partial exchange, 59%, 76%, and 84% exchange of ^{18}O for starting ^{16}O in the ferromagnetic superconductor $RuSr_2GdCu_2O_8$ (from D. J. Pringle et al., *Phys. Rev. B*, 1999-II, *59*, R11679–82). This figure was created using data from an inset in Fig. 2 in the original paper. The point for 0% exchange and 0 relative shift, which is common to all four frequencies, is off the scale of this truncated figure.

contribution (C_e) and an H-independent lattice contribution (C_l), an analysis of C(H) - C(9 T) also has the advantage that C_l cancels, leaving C_e(H), the contribution of greater interest. In the normal state C_e(H) = $\gamma_n T$, independent of H; in the mixed state C_e(H) includes a T-proportional term, γHT, and H-dependent terms; and in the superconducting state, C_e(0) = C_{es} [21].

In the quote above for temperatures between 1 and 50 K: γ_n corresponds to the coefficient of the normal-state electron contribution to the heat capacity, γ_n = 2.6 mJ/(mol•K^2); and 9 T corresponds to the second critical magnetic field, H_{c2} for MgB_2 [12]. Figure 12.4 shows details, as presented, of the low-temperature heat capacity for this type II superconductor.

Figure 12.4 illustrates the fact that the low-temperature heat capacity of $Mg^{11}B_2$ is more complex than the semiclassical model used. It will be a potentially major task to resolve the specific vibronic–electronic coupling that is the source of this complexity.

VORTEX LATTICES

Development of the theory of the vortex lattice arose from the 1950 paper by Ginzburg and Landau [1,2]. A solution to the Ginzburg–Landau equations was published in 1952 by Alexei Abrikosov [6,7]. Abrikosov's solution developed from an investigation of the consequences of having the Ginzburg–Landau parameter,

$$\kappa > \frac{1}{\sqrt{2}},$$

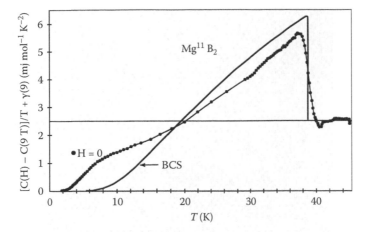

FIGURE 12.4 [C(H) – C(9 T)]/T v. T. The scale has been shifted by $\gamma(9\ T)$ to approximate $C_e(H)/T$ (see text quote above). The horizontal line represents $\gamma(9\ T)$. The error bars are ±0.1% of the total measured heat capacity. (From F. Bouquet, R. A. Fisher, N. E. Phillips, D. G. Hinks, and J. D. Jorgensen, *Phys. Rev. Letters*, 2001, *87*, 047001 1–4. adapted from Fig. 1. (a), p. 047001–2.)

when the surface energy of the superconductor is negative [7].* Under these conditions, a second-order phase transition occurs in the system that is experimentally marked by the lower critical magnetic field for type II superconductors, H_{c1}. The experimental and theoretical foundation for this second-order phase transition was first published in 1957 [23]. In this paper, Abrikosov reported the upper and lower critical magnetic fields, H_{c2} and H_{c1}, for alloys of Pb, with Tl and In. Abrikosov also compared experimental values for parameters with those obtained from the theory.

Vortex lattices predicted by the Abrikosov theory were first confirmed experimentally in 1967 by Essmann and Träeuble using electron microscopy imaging on samples of lead 4 atom% indium or niobium alloys; Figure 12.5 [24].

There are many images of superconducting vortex flux lattices available on the Internet. One of the sites that has been particularly informative and interesting is that at the Department of Physics, University of Oslo [26]. An early high-resolution image of a vortex flux lattice is that obtained by Hess and coworkers using scanning tunneling microscopy on a superconducting sample of niobium-diselenide, $NbSe_2$; Figure 12.6 [27].

Vortex lattices such as that shown in Figure 12.6 are essential to most of the commercial uses of superconductivity because of the significant increase in magnetic field that becomes available as a consequence of their presence.

Chapter 13 briefly returns to the five questions that were raised at the beginning of this book. It discusses some of the requirements for an integrated microscopic theory of superconductivity that will account for the experimental details of the subject

---

* Negative surface energy means that, under thermodynamic control, the surface will tend to expand at the expense of the bulk solid. This is a physical condition associated with a sufficiently high magnetic field in a type I superconductor. The problem was known early in the study of superconductivity. It attracted the attention of Landau and Ginzburg in their studies reported in 1950 [1].

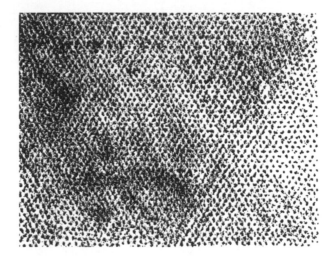

FIGURE 12.5   Electron microscope image of superconducting lead-4 atom% indium alloy, showing the presence of a triangular flux lattice as well as defects, T, 1.1 K (from U. Essmann, and H. Träuble, *Phys. Letters* 1967, *24A*, 526–7). This image is not identical to those presented in the original publication (from A. A. Abrikosov, *Rev. Mod. Phys.*, 2004, *76*, 975–9; Fig. 5, p. 978.)

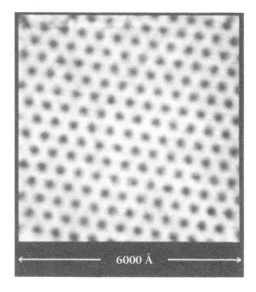

FIGURE 12.6   Abrikosov flux lattice produced by a 1 T magnetic field in $NbSe_2$ approximately 1×10^{-8} mho (black) to 1.5×10^{-9} mho (white). (From H. F. Hess, R. B. Robinson, R. C. Dynes, J. M. Valles, Jr., and J. V. Waszczak, *Phys. Rev. Letters*, 1989, *62*, 214–6; original Fig. 2, p. 215.)

from elemental superconductors to the exceptional detail observed in the heat capacity of type II superconductors.

REFERENCES

1. V. L. Ginsburg, and L. D. Landau, *Zh, Eksp. Theo. Fiz,*, 1950, *20*, 1064–84.
2. For an English translation of [1], see, D. ter Haar, Ed., *Collected Papers of L. D. Landau*, 1965, Gordon and Beach, London; pp. 546–568.
3. A. A. Abrikosov, *Dokl. Akad. Nauk SSSR*, 1952, *86*, 489.
4. A. A. Abrikosov, *Rev. Mod. Phys.*, 2004, *76*, 975–9.
5. L. Y. L. Shen, N. M. Senozan, and N. E. Phillips, *Phys. Rev. Letters*, 1965, *14*, 1025–7.
6. S. J. Williamson, *Phys. Rev. B*, 1970, *2*, 3545–56.
7. R. J. Trainor, M. B. Brodsky, *Phys. Rev. B*, 1975, *12*, 4867–9.
8. C. A. Reynolds, B. Serin, W. H. Wright, and L. B. Nesbitt, *Phys. Rev.*, 1950, *78*, 487.
9. E. Maxwell, *Phys. Rev.*, 1950, *78*, 477.
10. J. Bardeen, *Phys. Rev.*, 1950, *79*, 167–8.
11. J. Nagamatsu, N. Nakagawa, T. Muranaka, Y. Zenitani, and J. Akimitsu, *Nature*, 2001, *410*, 63–4.
12. S. L. Bud'ko, G. Lapertot, C. Petrovic, C. E. Cunningham, N. Anderson, and P. C. Canfield, *Phys. Rev.*, 2001, *86*, 1877–80.
13. M. Jones, and R. Marsh, *J. Am. Chem. Soc.*, 1954, *76*, 1434–6.
14. D. G. Hinks, H. Claus, and J. D. Jorgensen, *Nature*, 2001, *411*, 457–60.
15. *Ibid.,* data from, Figure 1, p. 458.
16. W. M. Haynes, Ed., *CRC Handbook of Chemistry and Physics*, 92nd edition, 2011, Taylor & Francis Group, Boca Raton, FL.
17. J. M. An, W. E. Pickett, *Phys. Rev. Letters*, 2001, *86*, 4366–9.
18. J. P. Franck, in *Physical Properties of High Temperature Superconductors IV*, D. M. Ginsberg, Editor, 1994, World Scientific Publishing, Singapore; pp. 189–293.
19. D. J. Pringle, J. L. Tallon, B. G. Walker, and H. J. Trodahl, *Phys. Rev. B*, 1999-II, *59*, R11679–82.
20. F. Bouquet, R. A. Fisher, N. E. Phillips, D. G. Hinks, and J. D. Jorgensen, *Phys. Rev. Letters*, 2001, *87*, 047001 1–4.
21. *Ibid.;* p. 047001-1, column 2, first full paragraph.
22. *Ibid.*; adapted from Fig. 1. (a), p. 047001–2.
23. A. A. Abrikosov, *Soviet Physics JETP*, 1957, *5*, 1174–82.
24. U. Essmann, and H. Träuble, *Phys. Letters* 1967, *24A*, 526–7.
25. Image source: A. A. Abrikosov, *Rev. Mod. Phys.*, 2004, *76*, 975–9; Fig. 5, p. 978.
26. [http://www.mn.uio.no/fysikk/english/research/groups/amks/superconductivity/sv/]
27. H. F. Hess, R. B. Robinson, R. C. Dynes, J. M. Valles, Jr., and J. V. Waszczak, *Phys. Rev. Letters*, 1989, *62*, 214–6.

13 Conclusions

WHAT ABOUT THE FIVE UNRESOLVED PROBLEMS IN SUPERCONDUCTIVITY?

The five areas that we introduced concerning questions related to the foundations for our views of superconductivity are listed, once again, below. Each question is followed by a statement of the understanding of the subject developed here that may provide a foundation for further development of theory or experiment in this area.

THE PATTERN OF ELEMENTAL SUPERCONDUCTORS AT ONE BAR IN THE PERIODIC TABLE

Table 13.1 presents the pattern of one-bar superconductors in the periodic table of the elements that we started with. The pattern in the superconducting critical temperature of the elements is complex. By use of the orbital angular momentum quantum number of the element and partial wave scattering theory, it has been possible to develop an understanding of the pattern including the ups and downs of the critical temperatures of superconductors in groups 2 through 14.* Elements in columns 3, 5, and 7 of the periodic table have significantly higher critical temperatures than elements in columns 4, 6, and 8 because of the low-temperature emergence of states with ns^1 basis conduction bands for the latter group of elements. Elements in the odd columns have odd numbers of electrons, and as metals can be conductors without electron promotion. This is the first group of robust superconductors in the table: La, V, Nb, Ta, Tc, and Re. Elements in even-numbered columns in this range can have closed subshell electronic ground states and require electron promotion to become metals. These metals have direct promotion access to ns^1 basis functions. Conduction bands with a basis of ns^1 atomic functions will exhibit partial wave scattering of electrons at the lowest attainable temperatures and end superconductivity. Superconductivity for these elements depends on the location of their open s-basis states in the overall density of states.

The second group of robust superconductors at one bar in the periodic table is located in columns 12, 13, and 14. This group includes Hg, In, Tl, Sn, and Pb. At low temperature, all of these elements have a closed d subshell. Mercury is the only member of the group that requires an open s subshell to form a superconductor. The question of bonding in the mercury superconducting state may be open for some time because of the serious difficulties associated with obtaining high-quality theoretical estimates for bonding states in atoms with 80 electrons each. Formation

---

* See Chapter 11.

TABLE 13.1

Periodic Table Showing the Bulk Elements that are Superconducting at One Atmosphere with Their Critical Temperatures, T_c

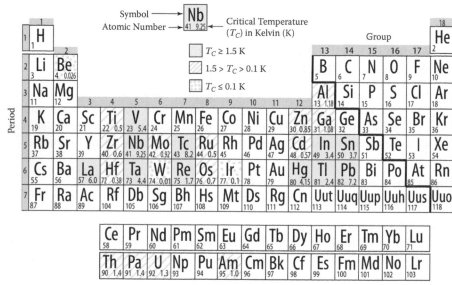

Note: Critical temperatures, T_c in kelvin, are shown at the bottom right for each element. Tungsten, W, is only superconducting in a film, not in bulk samples.

Source: C. Buzea, and K. Robbie, *Supercond. Sci. Techn*ol., 2005, *18*, R1–R8.

of diatomic mercury–mercury conduction components in the solid state is one of several possibilities. The mercurous ion system, Hg_2^{2+}, is well known in mercury chemistry, for example, calomel, Hg_2Cl_2. Calomel mercury is bonded with a mercury $6s$ σ-bond.

The patterns of one bar critical temperatures in the periodic table appear to be due to real features of atomic structure associated with the magnetic properties of electrons. These patterns need to be integrated into a revised, functional theory of superconductivity.

HIGH-TEMPERATURE SUPERCONDUCTIVITY, SUPERCONDUCTORS WITH CRITICAL TEMPERATURES, T_c > 77 K

High T_c is a lesson in the effect of occupied outer shells on the availability of inner shell electrons to form conduction bands. Copper metal will never be a superconductor because its $4s^1$ basis conduction band will have electronic scattering of conducting electrons at the lowest temperatures available. Planes of copper II ions are the heart of copper perovskite or cuprate-type superconductors such as the commercial yttrium barium copper oxide, YBCO. In these systems, the current flows in the copper II planes using $4d^1$ basis conducting orbitals.[*] The high critical temperature for cuprate superconductors is the result of the $4d$ basis conducting band; see Chapter 3, Table 8.3,[†] and the preceding discussion of Molecular Orbital Approaches to Resistivity. The d symmetry of the conduction band in cuprates and related high T_c superconductors has been experimentally confirmed many times. The d symmetry properties and the high T_c of the conduction band are not an accident. Rather, they are a consequence of the functional nature of superconductivity.

Unpaired Electron Spin in Superconductivity, For Example, the Knight Shift in Superconductors,[‡] and the Existence of Triplet Superconductors with Nonzero Electron Spin

All of the experimental data considered here is consistent with the view that the thin magnetic-field-penetrable layer at the surface of superconductors has significant electron spin at the lowest temperatures attainable. In fact, our understanding is that the data available from Knight shift experiments provides an accurate representation of the electron spin state on the interior of a functioning superconductor. This understanding depends on the results of the experiments that demonstrated that the electron spin state from Knight shift experiments and the superconductor electron spin state from Josephson junction experiments were identical for superconducting Sr_2RuO_4.[§]

The central question is: What processes are involved in the change in volume magnetic susceptibility at the superconducting critical temperature? The available data from electron spin states of Sr_2RuO_4 in the superconductor and in the London penetration layer suggests than no change of metal spin state is involved in the transition to superconductivity. Electron scattering from the conduction band is the experimental factor that dramatically changes at T_c. The role of this factor in magnetic susceptibility of metals is a prime topic for both experimental and theoretical inquiry. At this point, it appears that the end of dissipative electron scattering in the conduction band at T_c ends the definition of the Condon magnetic domains in the metal on the normal metal side of the phase transition. On the superconductor side of the transition, the end of dissipative electron scattering defines the superconductor body as a single magnetic domain. If an external magnetic field approaches or is present in the newly formed superconductor, it will generate an opposing diamagnetic current in the superconductor—Faraday's law of induction. Because resistivity in the bulk superconductor is zero, the induced current produces the perfect diamagnet.

No change of metal electron spin at the transition to superconductivity would avoid the problems associated with Mott transitions in systems with zero resistivity. Both magnetic-field- and electric-field-induced Mott transitions have been observed

---

[*] Only one of the $4d$ wave functions on any copper II ion is utilized as part of the conducting band for this superconductor. The other four $4d$ wave functions on each Cu II ion are all doubly occupied and nonconducting.

[†] See Table 8.3.

[‡] Knight shifts are a shift in nuclear magnetic resonance frequencies for nuclei that are caused by unpaired electron spin.

[§] See Chapter 4, Experimentally Verified Electron Spin in a Superconductor.

in the quantum Hall regime under conditions of zero resistance.[*] Though these phase transitions have appeared in the literature, we are not aware of literature discussions of the bonding and spin states in the insulator phase and the conductor phase in the quantum Hall regime for this type of phase transition. If we assume that the quantum Hall regime conducting phase is a spin paired state, a question arises concerning the distinction between this state and the insulating state associated with the Mott transition. If the conducting phase is a metal phase with zero resistance, the preceding question does not arise, at least not in the same difficult form.

Temperature below the threshold for partial wave scattering from the conduction band is the most reasonable cause for zero resistivity in the quantum Hall regime. This temperature is below the critical temperature for a superconductor of similar chemical composition. This condition is, in general, more stringent than the restrictions on the quantum Hall regime, as low levels of scattering are permissible for many quantum Hall effects; see Chapter 11.

Inadvertent Knight shift experiments in the quantum Hall regime appear to have been conducted by Andros and Knight during their early studies of the subject (see Chapter 4). These studies showed that in the quantum Hall regime[†] the electron spin state of conducting tin was that of a normal metal. In reality, all Knight shift experiments, because of the limited penetration depth of any active superconductor and temperatures below T_c, are conducted in what is effectively the quantum Hall regime.

HIGHER SPECIFIC HEAT CAPACITY AT CONSTANT PRESSURE IN SUPERCONDUCTORS AS COMPARED TO THE NORMAL METALS AT T_c COUPLED WITH MAGNETIC SUSCEPTIBILITY, $\chi = -1$,[‡] FOR THE SUPERCONDUCTING STATE

Data on the change in heat capacity at T_c for superconductors points to the scattering electrons in the conduction band having only a small effect, if any, on the heat capacity of the system. The higher heat capacity of superconductors at T_c as compared to the normal conductor at approximately the same temperature points to the thorough integration of the electronic and lattice wave functions in the superconducting state, including all of the particles in the system. This is clearest in the heat capacity of type II superconductors, which have heat capacities that significantly diverge from semiclassical models. The elimination of dissipative electron scattering at the critical temperature is a central contributor to the increased heat capacity of a superconductor.

---

[*] See Chapter 6, Table 6.2, and Chapter 10, Phase Transition Quanta in the Quantum Hall Regime.

[†] Tin layers in the Andros, Knight experiments were reported to be 40 Å thick; see Chapter 3: This depth is smaller than any penetration depth that has been reported in the literature for superconductors. The experimental temperatures ranged from above the superconducting critical temperature to approximately 1 K. Under these conditions, there would be no dissipative electron scattering in the tin sample; however, it would not be a superconductor.

[‡] Magnetic susceptibility, $\chi = -1.0$, has never been seen in normal, fully electron spin-paired material, such as cellulose, benzene, or countless other closed shell fully electron spin-paired materials, all of which are weakly diamagnetic. Many metals that are excellent conductors, including silver, are weakly diamagnetic, like spin-paired materials.

The abrupt change in the volume magnetic susceptibility, from the order of $\pm 10^{-5}$ to -1 associated with the change in electron scattering from the value associated with the resistivity of the normal conductor system just above T_c to zero for the superconductor. We are convinced that it is the zero scattering of conducting electrons that establishes the magnetic susceptibility that is essential to superconductivity, $\chi = -1$.

SUPERCONDUCTIVITY'S ALIGNMENT WITH THE REST OF ATOMIC AND MOLECULAR QUANTUM MECHANICS, FOR EXAMPLE, COMPATIBILITY OF SUPERCONDUCTIVITY WITH THE SOMMERFELD EQUATION AND THE QUANTUM MECHANICAL, ELECTRONIC FOUNDATIONS OF RESISTIVITY/CONDUCTANCE IN ORDINARY CIRCUITS

Sommerfeld's relationship accounted for the details of the Wiedemann–Franz Law. The relationship was central to the early adoption of quantum mechanics for atomic and molecular systems. Sommerfeld only utilized electronic wave functions for an electron gas, which suggests no essential involvement of lattice wave functions in the relationship. Use of partial wave scattering as a dominant route for electron scattering and resistance in metallic conductors meets this requirement and provides a conceptually clear guide to understanding and predicting superconductor critical temperatures.

Critical temperatures for the phase transition in superconductivity must intimately involve the presence or absence of occupied electronic wave functions that have a probability of scattering nonthermal electrons for a given temperature and pressure. This means that the critical temperature for a specific superconductor will depend on the temperature distribution of conducting states with $l = 0$ atomic basis functions. At temperatures low enough so that some states will not have partial wave scattering, if the density of $l = 0$ basis states becomes zero, a macroscopic 3D conductor will become a superconductor.

SUMMARY

In summary, we are confident that there is a new basis in modern quantum mechanics for developing a general and detailed understanding of superconducting phenomena. The questions that were raised at the start of the project will receive satisfactory answers generated within the fields of condensed matter physics and chemistry. We anticipate that the answers will follow the general outlines suggested below.

1. Superconducting phase transitions occur at a temperature determined by zero electronic scattering from the Fermi-level wave functions of the molecular electronic system. The critical temperature for a superconductor is determined by the temperature at which partial wave scattering from the conduction band becomes indistinguishably different from zero.
2. Superconductors must be three-dimensional metallic conductors, large enough to support a non-superconducting boundary layer.

3. A temperature-dependent, electronic energy gap must exist between the highest occupied conducting and lowest vacant bands of the superconductor.
4. Volume magnetic susceptibility for the superconductor must be −1.
5. High T_c superconductors follow the Landau–Ginzburg–Abrikosov formalism for magnetic field structure.

REFERENCES

1. C. Buzea, and K. Robbie, *Supercond. Sci. Technol.*, 2005, *18*, R1–R8.

Glossary

actinide: Any of the chemical elements with atomic numbers between 89, actinium, and 103, lawrencium, is an actinide. All of the elements with atomic numbers in the actinide range are radioactive.

adiabatic: Referring to a process with no heat transfer in or out.

allotrope: A phase of a material that has more than one phase at a given temperature and pressure. White and grey tin are allotropes. White tin: β-tin; negative Hall coefficient; conductor; superconductor; tetragonal, distorted face-centered cubic lattice. Grey tin: α-tin; positive Hall coefficient; semiconductor; low-temperature form; diamond lattice.

anion: A negatively charged ion that moves toward the anode, positive electrode, in an electrochemical experiment.

antiferromagnetism: Antiferromagnetic materials have quantum mechanically controlled order of electron spins on adjacent sub-lattice sites that are anti-parallel. See the schematic below.

antisymmetric: Referring to a wave function that changes sign on exchange of any two particles, electrons, whose wave properties are described by the wave function.

basis function: In molecular quantum mechanics, the basis functions are the atomic orbital mathematical functions complete with atomic quantum numbers that are used to create the linear combination of atomic orbitals, (LCAO), molecular orbitals, (MOs) used in calculations of energies and other properties.

Bose–Einstein statistics: Quantum statistics of indistinguishable particles, with wave functions that are symmetric under particle interchange, any number of which can occupy the same quantum state. Multiply occupied states of this type are referred to as Bose–Einstein condensates.

boson: A particle with a symmetric wave function that obeys Bose–Einstein statistics. A particle with an integer spin quantum number.

canonical variables: In physics, any pair of continuous variables whose Poisson brackets give a Kronecker delta, such as position and linear momentum. In quantum mechanics (e.g., Schrödinger's equation), electron position, q, and wave vector, k, are canonical variables. It is possible to solve Schrödinger's equation using either one of these variables, and you will obtain the same answer.

cation: A positively charged ion that moves toward the cathode, (negative electrode), in an electrochemical experiment.

calutron: Refers to a giant magnetic sector mass spectrometer used to separate weighable quantities of isotopically specific ions. The major use for calutrons to date has been the separation of radionuclides (e.g., isotopes of uranium for use in military devices).

closed shell: Electronic systems with an even number of electrons that are all spin paired and in their electronic ground state are referred to as a closed shell.

conduction band: When metallic elements bond together to form a solid metal to a first approximation, all of the bonds formed with a given global set of atomic basis functions will be degenerate and have the same energy. Because of the details of the quantum mechanics of these massively degenerate singly occupied wave functions their energies spread out into a "band" of conducting wave functions, that are nearly degenerate. There is a conduction band formed for each collection of valence level basis functions used to form bonds in the element.

continuity equation: A physics differential equation that describes transport of a conserved quantity. Many quantities are conserved in transport, for example, energy, momentum, mass, electric charge, and others. In the equation below the quantity is ϕ, ∇ is the gradient, • is the vector dot product, and f is the flux of the quantity ϕ.

$$\frac{\partial \phi}{\partial t} + \nabla \cdot f = 0$$

covalence: Langmuir defined covalence as the number of bonding electron pairs that one atom shares with other atoms. [1]

critical magnetic field: The magnetic field at which superconductivity ceases for a given temperature. In type II superconductors there is a lower critical magnetic field, H_{c1}, which defines the boundary between the pure superconducting phase and the mixed superconducting, normal metal phase—the vortex lattice. In type II superconductors, the maximum magnetic field for superconductivity is designated as H_{c2}.

critical temperature, superconducting: This is the temperature of the second-order phase transition between a bulk superconductor phase and a normal metallic conductor phase in the absence of external electromagnetic fields. The superconductor always exists at the low-temperature side of the phase boundary.

Curie temperature: Temperature above which a ferromagnet becomes a paramagnet, with a linear dependence of the reciprocal of magnetic susceptibility on temperature.

degree of freedom: One dimension for a particular motion. A translational degree of freedom is freedom to move along one of the axes in a three-dimensional system.

diamagnet: Diamagnets are nonmagnetic materials that have magnetic susceptibility <0. They have a tendency to form a magnetic field in opposition to an applied magnetic field. Pyrolytic carbon is an example of such a material.

Dirac delta: The Dirac delta, is defined as equal to zero for S(x), not equal to zero together with the requirement that the integral of the Dirac delta is equal to 1.

divergence: A differential mathematical operator, $\nabla\bullet$, which for a Cartesian vector field would be:

$$\nabla\bullet = \left(\hat{i}\,\frac{\partial}{\partial x} + \hat{j}\,\frac{\partial}{\partial y} + \hat{k}\,\frac{\partial}{\partial z} \right)\bullet$$

where $\hat{i}$, $\hat{j}$, and $\hat{k}$ are unit vectors in the x, y, and z axes, respectively.

ductility: The property of metals associated with the ability to draw metals into thin wires, deformability without loss of toughness.

electron effective mass: The apparent mass of the electron obtained by use of data from an electron cyclotron resonance experiment with the assumption that the electron momentum would match a thermal distribution for classical electron velocity. Magnetoresistance reduces electron momentum in these experiments, because of electron scattering, causing the effective change in electron mass to m_e^*. Values for electron effective mass have also been obtained from low-temperature heat capacity measurements.

electronegativity: Electron attracting power of an element. For the definition of Pauling electronegativity see Equation (7.1). Mulliken electronegativity is defined in Equation (7.2).

enthalpy: Referring to a system in thermodynamics, the sum of the system's internal energy and the product of the system volume times the pressure applied to it, generally referenced to an arbitrary zero.

entropy: A thermodynamic quantity introduced by Clausius in 1865 to describe the quantity of thermal energy in a system that is not available to do mechanical work. Entropy is also associated with randomness or disorder. Ludwig Boltzmann took his own life while on holiday in 1905. He had previously experienced severe periods of depression. His epitaph is the equation below. S is entropy. k is Boltzmann's constant. W is the number of possible microscopic states that could correspond to the macroscopic state of the system. log refers to the natural logarithm currently represented as ln.

$$S = k \log W$$

equipartition theorem: A short statement of this classical theorem in statistical mechanics is that in a mechanical system, independent particles at thermal equilibrium have $\tfrac{1}{2}k_BT$ of thermal energy on average for each degree of freedom.

Fermi contact: An informal description of the fact that in partial wave scattering involving s-basis wave functions, it is possible for the electron and nucleus to adopt the same geometric coordinates with a certain probability. This "contact" interaction between an electron and nucleus makes exchange of energy, linear momentum, and angular momentum (spin) between the two particles possible.

Fermi–Dirac statistics: Quantum statistics of identical particles with wave functions that are antisymmetric with reference to exchange of coordinates of any two particles.

Fermi level: Fermi level electrons are electrons in the highest energy occupied state of a metal at temperature approaching 0 K and one bar pressure.

fermion: A particle with an antisymmetric wave function that obeys Fermi–Dirac statistics. A particle with a half-integer spin quantum number. Particles with integer spin quantum numbers are called *bosons*. The two quantum classes have many fundamentally different properties.

Fermi surface: A surface in momentum space, *k* space, for the momentum of the highest energy electrons in the system at very low temperatures, ~0 K.

ferrimagnetism: Ferrimagnetic materials are ordered solid-state spin systems that are intermediate between ferromagnetic and antiferromagnetic order. That is, the spins in a given subdomain are only partially parallel (see the schematic for spin order in a hypothetical ferrimagnet below).

ferromagnetism: Local, quantum mechanically controlled, alignment of unpaired electron spin dipoles in the solid state. This is the fundamental mechanism for formation of permanent magnets in metals analogous to iron. See schematic, below.

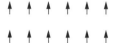

Gaussian units: A system of electromagnetic units that uses the cgs metric for space, time and mass. The treatment of electromagnetic fields is distinct in Gaussian and SI units because of the differences that arise in the Maxwell equations in the two unit systems.

graphene: A single monolayer of hexagonally bonded carbon atoms that has been produced either by growth of a carbon monolayer under controlled conditions or by pealing a single layer from graphite using plastic tape. [2]

heteronuclear: A diatomic bond having two different nuclei (e.g., cesium fluoride, CsF).

homonuclear: A diatomic bond having only one kind of atomic nucleus (e.g., nitrogen gas, N_2).

Hund's rule: Friedrich Hund, German physicist, developed three empirical rules for the terms in atomic spectroscopy in the mid-1920s. His rule for maximum multiplicity states: For a given electron configuration (number of *s*, *p*, *d*, and *f* electrons in an atom), the term with maximum electron spin multiplicity has the lowest energy.

hysteresis: A property whose value depends upon the direction of approach (e.g., the magnetic field in an iron core electromagnet as a function of rising or falling magnet current). The area of the closed hysteresis curve is a measure of the energy lost in the cycle.

intermetallic: Intermetallic compounds are metal alloys that have a definite chemical structure and formula, such as, $CeIrIn_5$.

isotopomer: An isomer produced by isotopic substitution (e.g., $^{24}Mg^{11}B_2$ and $^{26}Mg^{10}B_2$). It is also possible to have stereoisomeric materials where the stereoisomerism is due to isotopic substitution. For example, see the optical isotopomers of ethanol-1-d, CH3CHDOH below. The designations R and S refer to the absolute orientation in space of the four different groups attached to the central carbon, represented by the intersection of the four bonds, in the drawing.

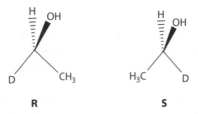

R   **S**

k **space:** Electron momentum space.

Laplacian operator: The differential operator, ∇^2, "del squared" from Laplace's equation, which in two dimensions is:

$$\nabla^2\phi = \frac{\partial^2\phi}{\partial x^2} + \frac{\partial^2\phi}{\partial y^2} = 0$$

lanthanide: Any of the chemical elements between lanthanum, atomic number 57, and lutetium, atomic number 71. A rare earth or *f* transition metal.

lattice parameter: There can be up to three lattice parameters for a crystal, a_0. They give the distance between unit cells either in pm, 10^{-12} m or Å, 10^{-8} cm.

"localized" bond: Chemists often speak of nonconducting electron pair bonds as "localized." Photoelectron spectra show that the electrons that form these bonds have the bulk of their electron density associated with the two "bonded" atoms. The concept of "localization" is the basis for the ball and stick models of chemical structure. Such models are not to be taken literally as far as electron localization is concerned. The models have, nonetheless been very useful in developing understanding of chemical structure.

magnetic permeability: Magnetic permeability, μ, is the relationship between magnetic induction, B, and magnetic field, H, in a linear magnetic medium, $B = \mu H$.

magnetic susceptibility: Magnetic volume susceptibility, χ_v, is a tensor, and is defined by the partial derivative of the magnetization, M, with reference to the magnetizing field, H,

$$\chi = \frac{\delta M}{\delta H}$$

It is dimensionless in SI and cgs units. To convert to SI units, multiply the cgs value by 4π.

mesoscopic: Referring to a size scale between micro and macroscopic.

molecular orbital: A wave function that extends over a molecule. In a molecular orbital sense a metal single crystal is a mesoscopic macromolecule.

normalized: Referring to a wave function that describes the wave properties of an electron. If the probability of finding the electron somewhere in space is 1, the wave function is said to be normalized. The integral definition for a molecular wave function, $\psi(x)$ is

$$\int_{-\infty}^{\infty} \Psi^*(x)\Psi(x)\,dx = 1$$

open shell: Referring to atomic or molecular electronic systems that have at least one unpaired electron. Electronic systems with an even number of electrons that are all spin paired and in their electronic ground state are referred to as closed shell.

orthogonal: In molecular quantum mechanics referring to two orbitals in the same molecule, if the all-space integral of the product of the two orbitals is zero, they are orthogonal.

orbital: Wave function for an electron or pair of electrons associated with one or more atomic cores. [3]

paramagnet: A nonmagnetic material that is attracted to magnets, for example, oxygen, O_2, in the air. The two unpaired electrons in the oxygen molecule have a magnetic moment, so the gas is attracted to the poles of an electromagnet.

parity: A parity transformation, or parity inversion, is the change in sign of one spatial coordinate in three dimensions. If the system is unchanged on inversion it has even parity. If the variable sign changes on inversion it has odd parity.

penetration depth: Width of the surface layer of a superconductor that can be penetrated by an external magnetic field that is weaker than the critical magnetic field at the observation temperature. Also called the London penetration depth.

permeability: Magnetic permeability of a metal is the ratio of the magnetic induction, B, in a metal to the applied magnetic field, H.

$$B = \mu H$$

phonon: Vibrational energy in solids. Normal mode vibrations in a solid have electromagnetic quanta that are analogous to vibrations in molecules. Acoustic phonons in solids correspond to the sound energy associated with sound waves in the medium. For electromagnetic phonons in metals the vibrational frequencies are in the 10^{12} s^{-1} range.

π-bond: A chemical bond formed by edge-on overlap of two generally $2p$-basis functions. Increased electron density between the two bonded atoms is found above and below the plane that contains the radial nodes of the two $2p$-basis functions. These bonds generally have lower bond enthalpy than σ-bonds between the same atoms.

π-orbital: A molecular wave function for a bonding or nonbonding electron that has a nodal symmetry plane that connects adjacent bonded atoms that are the source of the basis functions for the orbital. The π-orbital that lies above and below the plane of monolayer graphene is an archetypical example in condensed matter physics.

plasmon: A plasmon is a collective quantum oscillation of valence electrons of a metal. Plasmons are responsible for many of the optical properties of metals including color, luster, and sheen.

pnictide: Compounds formed between cations and members of the nitrogen family. Iron pnictides are compounds of iron with members of the nitrogen family, N, P, As, e.g., the superconductor $CeO_{1-x}F_xFeAs$. [3]

Poisson bracket: A partial differential operator that is important in Hamiltonian mechanics, and canonical variables, defined by the following equation for the Poisson bracket of distance, x, and momentum, k. The canonical coordinates qi and pi in the equation are another pair of canonical coordinates.

$$[x,k] = \sum_{i=1}^{N} \left[\frac{\partial x}{\partial q_i} \frac{\partial k}{\partial p_i} - \frac{\partial x}{\partial p_i} \frac{\partial k}{\partial q_i} \right]$$

promotion energy: The promotion energy is the amount of energy that must be put into an atom in its electronic ground state to produce the bonded electronic configuration of the element. In beryllium, for example, the atomic ground state configuration is $1s^2 2s^2$. This electronic configuration is neither bonding nor would it be conducting. To form beryllium metal it is necessary to promote one $2s$ electron to a $2p$ wave function and create the open shell configuration $1s^2 2s^1 2p^1$. In this electronic configuration the atoms can form metallic bonds and the bonded metal will conduct electricity. The promotion energy is the difference in energy between the two electronic configurations for a single atom.

refractory metal: One of a group of metal elements that are unusually high melting and wear resistant. This group includes metals in columns 4 through 9 of the periodic table, with the exception of Mn, Fe, Co, and Tc. It seems likely that some of the members of the actinide series would be included in this group, were it not for their radioactivity.

resistivity:

$$\rho = R \frac{A}{L},$$

resistivity, ρ (Ω-m) is the product of resistance, R (Ω) and area of the conductor, A (m^2), divided by its length, L. Electrical resistivity is a consequence of the scattering of conduction electrons.

scalar: Referring to a quantity having only magnitude, not direction.

self-consistent field: The minimum energy field that is obtained by application of the variational principle in SCF molecular orbital calculations.

σ-bond: A bond between two atoms that has radially symmetric electron density, to first order, along the bond axis.

σ-orbital: A molecular wave function for bonding or nonbonding electrons that has radially symmetric electron density with reference to the bond axis of the atoms that are the source of the basis functions for the orbital.

SI units: An internationally recognized system that has been adopted worldwide, with the exception of the United States and one or two other countries. SI units uses the mks metric for space, time, and mass. The treatment of electromagnetic fields is distinct in SI and Gaussian units because of the differences that arise in the Maxwell equations in the two unit systems.

specific conductance: The reciprocal of resistivity. The conductivity per unit length of a material of unit cross-sectional area. The SI unit for specific conductance is Sm^{-1}, siemens per meter.

specific resistance: The resistivity of a material. The resistance per unit length of a material of unit cross-sectional area. In SI units resistivity is given in Ωm.

Stern–Gerlach: A 1922 experiment in quantum mechanics that used a magnetic field to separate silver atoms based on atomic electron angular momentum states.

susceptibility: Volume susceptibility, also called derivative susceptibility is defined as the partial derivative of the magnetization, M, with reference to the applied magnetic field, H. The definition is valid only in the range where the relationship below is linear.

$$\chi = \frac{\delta M}{\delta H}$$

superconductivity: Superconductivity is the capacity of a bulk material to form a current generated magnetic field that is indefinitely persistent without energy input into the current.

tautology: Repetition of a word or idea in the same or other words.

tensor: Tensors are mathematic abstract objects that contain an array of components that are functions of coordinates such that, under a transformation of coordinates, the new components are directly related in a defined way to the transformation and to the original components. The order of a tensor tells you the number of components. Zero order tensors have a single component and are scalars. First-order tensors in n-dimensional space have the number of components corresponding to the coordinate system and are vectors. Second-order tensors in a four-dimensional coordinate system have 16 components. In general, second-order tensors in a n-dimensional coordinate system have n^2 components. You can think of a second rank tensor as a matrix of vectors in the coordinate system.

term symbol: An abbreviated quantum mechanical (Russell-Saunders) description of the angular momentum quantum numbers in a many electron atom: $^{2s+1}L_J$. Here, $(2s + 1)$ is the electron spin multiplicity. L, is the total orbital quantum number in spectroscopic notation, and J is the total angular momentum quantum number.

triplet: An electronic spin state involving two parallel spin electrons. Triplet states have three spectroscopic energy levels.

unit cell: The smallest set of atoms in a crystal that contains the crystal symmetry. Repetition of the unit cell in three dimensions will reproduce the crystal.

valence: Bonding capacity of an atom, atomic valence. The valence of carbon is four. In general, carbon atoms form four two electron bonds with other atoms.

valence level: Referring to the highest principal quantum number for an atom. Bonds to an atom generally involve valence level electrons.

variational principle: This principle states that for any normalized molecular wave function, ψ, the expectation value of the Hamiltonian for ψ must be greater than or equal to the ground state energy. In the LCAO MO approximation the atomic basis function coefficients are varied to minimize the total energy given by the molecular Hamiltonian.

vortex lattice: In type II superconductors there is a maximum magnetic field for observation of superconductivity with a type I magnetic structure. The lower critical field, H_{c1}, marks the end of this region and the beginning of the vortex lattice which adopts a mixed superconductor—normal metal structure to maximize the magnetic field and minimize the free energy of the system.

wave vector: A canonical variable in quantum mechanics, k. An electron wave vector, k, is the vector whose magnitude is $2\pi/\lambda$ (radians/m), where λ is the wavelength of the electron in meters. Wave vector is also defined as $p/\hbar$, p is electron momentum and $\hbar$ is Planck's constant divided by 2π.

work function: Minimum energy, eV, required to remove an electron from a metal solid to a stable point outside the surface. The work function measures the energy of the Fermi level under a given set of conditions.

Young's modulus: The ratio of the tensile stress to the tensile strain for an elastic object. It is the force per unit area divided by the change in length per unit length for a stretched object.

REFERENCES

1. Langmuir, Proc. Nat. Acad. Sci., 1919. 5, 252–9.
2. A.K. Geim, K. S. Novoselov, Nature Materials, 2007, 6, 183–91.
3. Y. Kamihara, T. Watanabe, M. Hirano, and H. Hosono, J. Ameri. Chem. Soc., 2008, 130, 3296-7.
3. R. S. Mulliken, *Phys. Rev.*, 1932, *41*, 50.

Variables, Constants, Acronyms

κ: Thermal conductivity, W/(m•K)

λ: Wavelength, m

λ_e : Wavelength of the electron, m

μ: Magnetic permeability, H/m, H, henry

μ_0 : Permeability of free space, $4\pi \times 10^{-7}$ H/m, H, henry

ρ: Resistivity, Ω•m

σ: Electrical conductivity, S/m, S, siemens

χ_v: Volume susceptibility (dimensionless)

Ω: Ohm, SI unit of resistance

amu: Atomic mass unit, 1/12 the mass of one atom of carbon-12, ^{12}C, the SI unit for atomic mass

B: Magnetic induction, magnetic flux density, T, tesla, 10000 gauss

C: Coulomb, SI unit of electric charge

cgs: Centimeter–gram–second units used for mechanics and extended to cover electromagnetism in the Gaussian system of units. Largely supplanted by SI units

dHvA: de Haas (and) van Alphen, discoverers of oscillations in magnetization with magnetic field changes or other perturbations to magnetoresistance

e: Electron charge, $1.60217646•10^{-19}$ coulomb

e^2/h: Magnetoconductance quantum, S, siemens, Ω^{-1}

$2e^2/h$: Electrical conductance quantum, S, siemens, Ω^{-1}

EPR: A. Einstein, B. Podolsky, N. Rosen, authors of a well-known seeming paradox in quantum mechanics. (See . Einstein, B. Podolsky, N. Rosen, *Phys. Rev.*, 1935, *47*, 77780.)

epr: Electron paramagnetic resonance

FET: Field effect transistor

H: Henry, SI unit of magnetic inductance

H: Magnetizing field, A/m

H_c : Critical magnetic field at a given temperature for a type I superconductor. The maximum magnetic field at T for a type I superconductor

H_{c1} : First (lower) critical field for type II superconductors that occurs at the boundary of superconducting and the mixed superconducting and normal state, the vortex lattice

H_{c2} : Second (upper) critical field for a type II superconductor. The magnetic field at T above which superconductivity is not observed in a type II superconductor

h: Planck's constant, $6.626068•10^{-34}$ J•s

$\hbar$: Reduced Planck's constant, h-bar, Planck's constant divided by 2π

h/e^2: Magnetoresistance quantum, Ω

$h/2e^2$: Electric resistance quantum, Ω

$h/2m_e$: Quantum of circulation, m^2/s

h/e: Magnetic flux quantum, Wb, weber, V•s

K: Kelvin, temperature

k: Wave vector, $p/\hbar$, radians/m

k_B : Boltzmann constant, J/K

l: Electron angular momentum quantum number, also called azimuthal quantum number (dimensionless)

LCAO: Linear combination of atomic orbitals

M: Magnetization, Am^{-1}

MOS: Metal oxide semiconductor also metal oxide silicon

MRI: Magnetic resonance imaging

m: Magnetic quantum number (unitless)

m_e: Electron mass, $9.10938188 \cdot 10^{-31}$ kg

m_e^*: Electron "effective mass"

m_s: z-component of the electron spin quantum number (unitless)

mks: A set of space, mass, and time units based on meter, kilogram, and second. The base unit metric for the SI system of units

n: Carrier density in a conductor (number of charge carriers/atom)

n: Principal quantum number (unitless)

NMR: Nuclear magnetic resonance

ppm: Parts per million

QHE: Quantum Hall Effect

r_0: Range of scattering potential, m

R_H: Hall coefficient

s: Electron spin quantum number (unitless)

s: Second, SI unit of time

S: Siemens, SI unit of conductance, Ω^{-1}

SCF: Self-consistent field, the main procedure minimizing electronic energies in MO calculations.

SI: Système International (French), the international system of units

SQUID: Superconducting quantum interference device, a quantum magnetometer based on Josephson junctions

T: Temperature, K, kelvin

T_c : Superconductor critical temperature, K

T: Tesla, SI unit of magnetic flux density, equal to 10,000 gauss

V: Electric potential energy, SI unit volt

Index

Page numbers followed by f indicate figure
Page numbers followed by t indicate table